THE DIGITAL ELECTRONIC WATCH

THE DIGITAL ELECTRONIC WATCH

Tom M. Hyltin

President, Micro Display Systems, Inc.

Dallas, Texas

VNR

VAN NOSTRAND REINHOLD COMPANY

NEW YORK CINCINNATI ATLANTA DALLAS SAN FRANCISCO

LONDON TORONTO MELBOURNE

Van Nostrand Reinhold Company Regional Offices:
New York Cincinnati Atlanta Dallas San Francisco

Van Nostrand Reinhold Company International Offices:
London Toronto Melbourne

Library of Congress Catalog Card Number: 78-1620
ISBN : 0-442-22596-2

Manufactured in the United States of America

Published by Van Nostrand Reinhold Company
135 West 50th Street, New York, N.Y. 10019

Published simultaneously in Canada by Van Nostrand Reinhold Ltd.

15 14 13 12 11 10 9 8 7 6 5 4 3 2 1

Library of Congress Cataloging in Publication Data

Hyltin, Tom M
The digital electronic watch.

Includes index.
1. Electronic watches, Digital. I. Title.
TS544.H93 681'.114 78-1620
ISBN 0-442-22596-2

Preface

A new "kid" moved into the neighborhood, and there followed a period of "sizing-up" and getting to know "him." The kid is the digital electronic watch, and it moved into a very mature mechanical watch industry. Those in the watch industry were talking LED, LCD and chip without adequate understanding, while those in the electronic field were having to learn about faceted crystals and case procuring.

The purpose of this book is to provide a background for everyone in the watch industry—helping them to get to know the "new kid in the neighborhood." Although engineers and manufacturers may view this a reference book, its real purpose is to provide information for the jeweler, store buyer, salesman, and watchmaker that will allow them to do their jobs better.

Chapter 1 through Chapter 8 are an explanation of the various components that go to make a digital watch—the basic information needed to understand these watches. Chapters 9 and 10 discuss servicing—not just how to fix, but setting up the bench, and service philosophy. The Appendixes of the book include a reference section—for both the jeweler and the watchmaker—as well as a section to show how the watch operates and what is the setting procedure. Also included is information on special features and battery replacement.

The author wishes to express his appreciation for the encouragement given by Mr. Harold Herman of the American Watchmakers Institute. The data contained in the Appendixes are summarized from material assembled by the AWI for a digital watch course and without their kind permission, this section would not have been possible.

The people at Micro Display Systems have also contributed their time to this effort. Photographs and test data were prepared by Gary Schneider, some line drawings were made by Gary Vaughn and Tom Conn, and the manuscript was typed, several times, by Penny Morgan. Valuable technical discussions were also held with Dr. Robert H. Schnurr and Robert C. Jones. Display samples were supplied and helpful discussion on LED watches and technology were held with Bob Cole and Carrole Smith of Texas Instruments. Their assitance is appreciated.

TOM M. HYLTIN

Contents

THE DIGITAL ELECTRONIC WATCH

1

Creating the Digital Electronic Watch

PEOPLE AND HISTORY

How did the digital watch get started? Who was the inventor? The story of the digital watch is about people, who believed that a complete electronic computer, its power source, and a digital readout could be put into a package the size of two half dollar pieces.

The digital watch couldn't have been made with vacuum tubes or transistors. But the integrated circuit (IC), as invented by Jack Kilby at Texas Instruments and in a different form by Bob Noyce at Fairchild, offered the compact electronic circuitry that was required. Several years after this invention, Patrick E. Haggerty, President of Texas Instruments, wrote of the "pervasiveness of electronics."* In this paper, he forecasted that in the years to come, integrated electronics would become a part of every person's life and being. The digital wristwatch is only one of the endless variety of products that has resulted from the semiconductor revolution of the 1960s.

Before there could be a watch, there had to be a compact, low-power display. In 1961, as the basic work on the development of ICs was in progress, Bob Biard and Gary Pittman at Texas Instruments first detected light (in the infrared region) being given off by a gallium arsenide diode. The following year, Holonyak and Bevacqua working at General Electric Company were able to modify the diode structure so that light was given off in the visible region. Production availability of these devices had to await a lot of work "in the trenches" as advances

* *Proceedings of the I.R.E.*, Vol. 52, No. 12, 1400–05, (1964).

were made in the ability to manufacture complex gallium arsenide devices.

At the same time, George Heilmeier and his associates at RCA were experimenting with liquid crystal devices. Liquid crystals have been known since 1888 and were first used in controlling the flow of light in 1936. Liquid crystal displays (LCDs) of the dynamic scattering type were described in a paper by the Heilmeier group in 1967. This type of display was used in early liquid crystal watches, but it still required higher voltages than was convenient in a watch. The final display step was taken by James Ferguson at Kent State University in 1971 with the development of the twisted nematic field effect display. This is a type of display where the polarization of light is twisted in order to achieve on/off operation at very low applied voltage.

This was the technical environment when the digital electronic watch was created. It was exciting—like living through the industrial revolution. The excitement and opportunities were recognized broadly. These people, working independently, contributed to the creation of the digital electronic watch.

John Bergey, an engineer at Hamilton Watch Company in Lancaster, Pennsylvania was doing research toward making a digital electronic watch. His thorough assessment of the available technologies resulted in patents for Mr. Bergey and for the Hamilton Watch Company which charted the course for the digital electronic watch for the decade following his patent applications in 1967 and 1968.

George Theiss, president of a small electronics company in a suburb of Dallas, first saw an LED display in 1966 and instantly realized that this was the display needed to build the digital electronic watch. As his company grew, he was able to add another talented engineer, Willy Crabtree. Resulting from the collaboration of these two engineers, working independently of John Bergey, was the world's first digital wrist watch. After a meeting with Bergey of Hamilton Watch Company, this became the first watch sold under the PULSAR name. The world of timekeeping has never and will never be the same again.

THE PROBLEM OF SIZE

Throughout the 400-year history of mechanical timekeeping, there has been a tendency toward size reduction. A reduction from big clocks to small clocks, small clocks to pocket watches, pocket watches to large men's wristwatches, and on down to the tiny watches used by ladies in rings, etc. The digital electronic watch is a product of the 1970s, and had to closely match the size of a watch desired by the people of the 1970s. This was one of the larger challenges faced by the electronics designers when digital watch engineering began.

Figure 1-1 gives an indication of the volume within the watch case, based upon the case diameter. The upper line shows the volume contained within the case and crystal. The lower line shows the volume between the dial and the back of the case. The volume available in a watch case goes from 0.5 cubic centimeter (cc) up to 9 or 10 cubic centimeters, depending upon how large a case you wish to consider as the

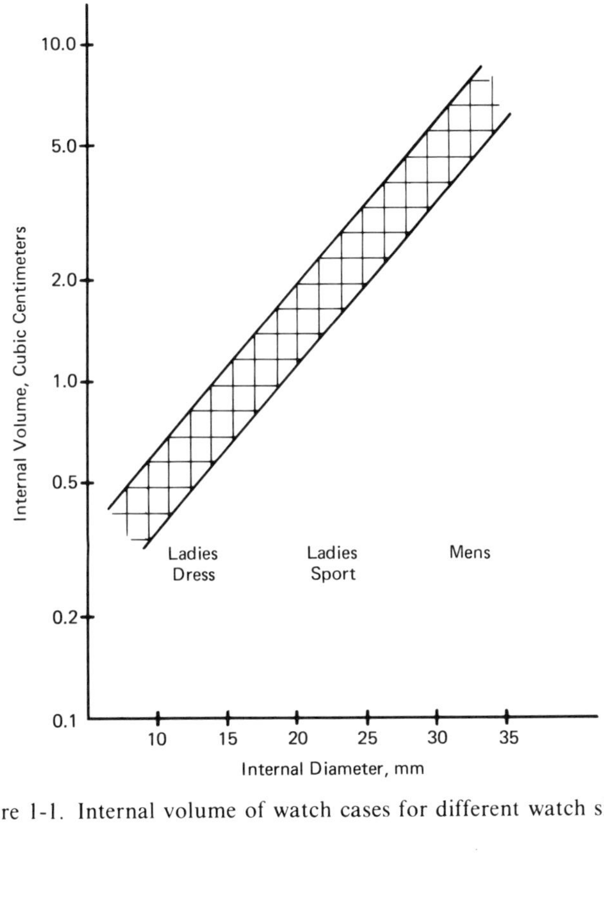

Figure 1-1. Internal volume of watch cases for different watch sizes.

upper end. For comparison, if the average thimble is filled with water, the amount of water in the thimble is about 3 cubic centimeters. To do a nicely styled ladies watch, the module can be not more than 1 cubic centimeter, or about one-third the volume of the average thimble. To provide a frequency standard, integrated circuit, display, and power source in a volume this size was truly a worthy challenge.

The first watches, both light emitting diode and liquid crystal displays, were not very efficient in their use of either space or power. The technical barriers to overcome in making the watch were such that the niceties (low-power consumption and small size) had to await future generations of digital electronic watches. Nowhere are the technical problems more apparent than in one of the first digital watches built by George Theiss and Willy Crabtree. Figure 1-2 shows the case and parts for a watch put together by Electro-Data for Hamilton (this was prior to the use of the PULSAR name). At the left is the case with one push button. To the right of the case are the two ceramic circuit carriers that contain all of the electronic circuitry of the watch. The upper of the two contains the display and the display driver integrated circuit. This consists of six ceramic circuit boards (substrates); the upper four substrates have the digits that show hours and minutes, and the lower two substrates have the digits that show seconds. The dark rectangle on each ceramic substrate is the integrated circuit that drives that digit.

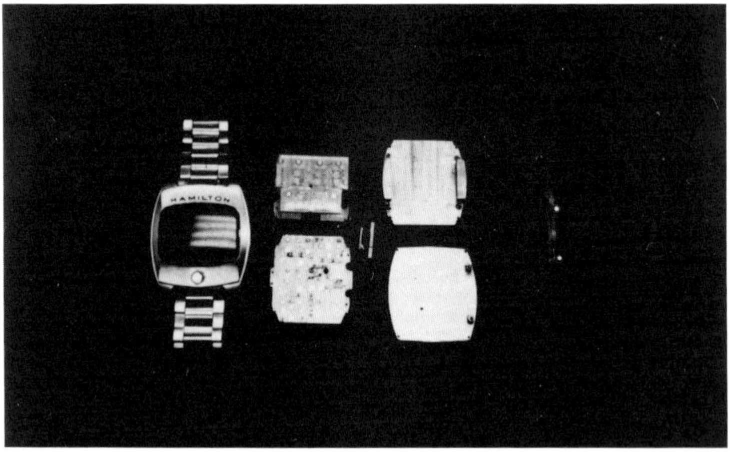

Figure 1-2. Early digital watch. Manufactured by Electro-Data.

Shown below the display carrier is the logic carrier that contains all functions from quartz oscillation to the calculation of time. When the watch is assembled, the logic carrier is mounted back-to-back with the display carrier. This carrier consists of nine ceramic substrates containing the integrated circuits that compute time and operate the watch. In the entire watch, there are over 40 integrated circuits. Next to the ceramic substrates is shown the unpackaged quartz that is used as the frequency standard in the watch. The two metal pieces provide spacing and backing plates for the watch. On the right is shown the rechargable silver battery which operated the watch. One charge was good for one or two months operation.

Figure 1-3 shows the electronics mounted into the watch case. The quartz crystal bar is shown against the left hand edge of the watch case. This photo also shows 34 of the integrated circuits used in the watch. The total volume occupied by the electronics and battery was about 10 cc.

Certainly not the first LCD watch, but an early effort in a complex liquid crystal display watch, was the watch developed by Texas Instru-

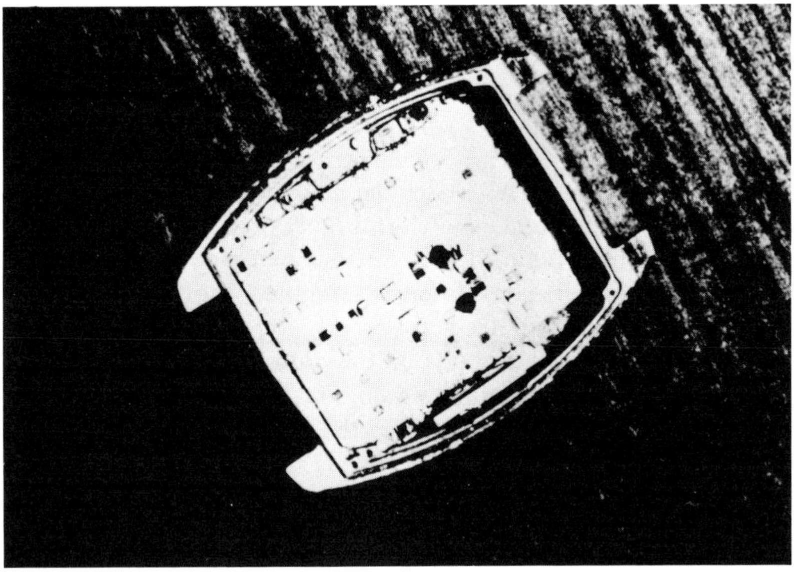

Figure 1-3. Assembled early digital watch. Manufactured by Electro-Data.

Figure 1-4. The module of a watch. Developed by Texas Instruments for Ebauches S.A.

ments for Ebauches S.A., under the guidance of Bob Schnurr. This watch followed the LED shown in Figures 1-2 and 1-3 by approximately 24 months. Figure 1-4 shows the front view of the watch case and display configuration chosen by Longines. The date was shown on top, the hours and minutes across the center (this illustration shows 24-hour time) and the seconds on the bottom. All push buttons for setting were on the back of the watch. Figure 1-5 is the rear view of the module showing the two batteries, the quartz, and the black items, which comprise the transformer. The ceramic substrate beneath the batteries is shown in Figure 1-6. The transformer is again the black object on the top side of the ceramic. The two black objects below it are the epoxy used to encapsulate the integrated circuits. The trimmer capacitor is located between the integrated circuits, and 51 contacts to the liquid crystal display are located around the outside edge. Four

Figure 1-5. Rear view of the module of a Texas Instruments Watch for Ebauches S.A.

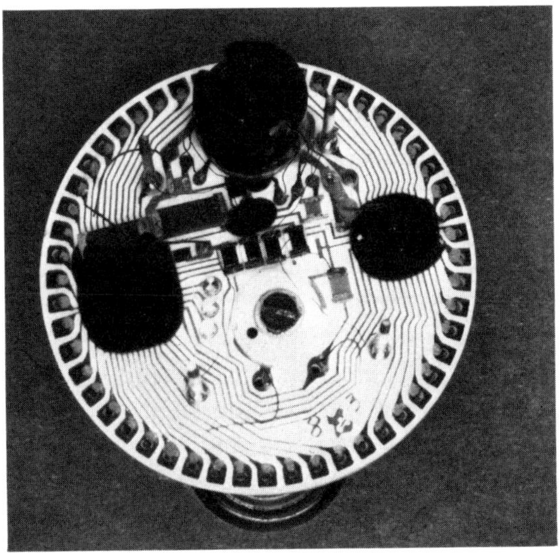

Figure 1-6. The ceramic substrate of a watch. Developed by Texas Instruments for Ebauches S.A.

posts coming from the ceramic substrate make contact with the push buttons on the back of the case.

Neither of the two watches shown in Figures 1-2 through 1-6 were built in production quantities. That step awaited more mature technology in the displays and integrated circuits that contribute to the digital electronic watch.

A FEW TECHNICAL TERMS

Prior to the description of the parts that make up the modern digital electronic watch, there should be some explanation of words not encountered in everyday conversation.

Module	The equivalent of a mechanical watch movement (or ebauche). It is the collection of displays, circuits, and other parts needed to keep and display time. The shape is such that it can be put into a watch case.
Digital	Showing time by use of numbers, not the position of hands. Both mechanical and electronic watches can have digital display of time, but common usage of digital is now implying all electronic.
Electronic	Operation and powering of a watch based upon the flow of electricity.
Solid-State	A term applied to equipment where all operations of that equipment are a result of transistors or integrated circuits.
Integrated Circuit	A tiny electronic circuit that contains hundreds or thousands of transistors and their associated components, which together perform a given job. A description of the integrated circuit doing the electronic watch job is given in Chapter 3.
Display	The part of the solid-state digital watch that shows letters and numbers. This can be LED (Light Emitting Diode) or LCD (Liquid Crystal Display). These display technologies are discussed further in Chapters 5 and 7.
Alphanumeric	A new word which means that a display can

form both the alphabet (alpha) and numbers (numeric). In a watch, it implies the capability to show day of the week and perhaps month in the same place that the numbers showing time generally appear.

THE REQUIREMENTS FOR A DIGITAL ELECTRONIC WATCH

All electronic watches share certain features which are necessary for a digital electronic watch to function. Figure 1-7 shows these components and their interconnection (in electronics, it is referred to as a block diagram). There is the oscillator frequency standard and the integrated circuit, which powers and is controlled by the oscillating element. The IC provides the frequency of oscillation down to that which can be used for timekeeping, normally one pulse per second. After a rate of one pulse per second is achieved, the IC does the necessary counting and decoding of the information so that the proper digits are produced by the display. In the case of liquid crystal displays, the IC drives the display directly. In many of the LEDs, that have been manufactured, the current required by the display is more than can be conveniently supplied by the IC. Therefore, outputs go through an output driver and then to the display. A battery powers the IC. There is a switch interface to the case for calling different functions and setting.

There is a direct correlation between the operation of an ordinary mechanical watch, a tuning fork type of watch (such as the Accutron®),

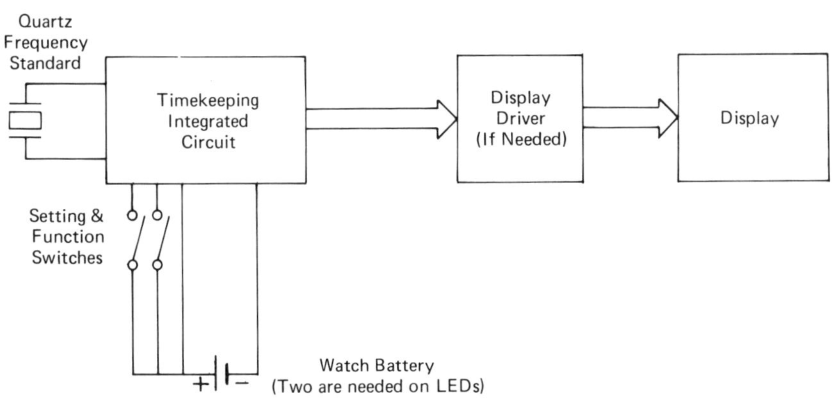

Figure 1-7. Block diagram of the components and their interconnection in a digital electronic watch.

Table 1-1. Comparison of Mechanical, Tuning Fork, Analog Quartz, and Digital Electronic Watches.

	Oscillating Element	Frequency Divider	Hours, Minutes, Seconds	Display	Power Source
Mechanical watch	Balance wheel	Timing chain (gears)	Dial chain (gears)	Hands, dial	Mainspring
Tuning fork watch	Tuning fork	Ratchet wheel	Dial chain (gears)	Hands, dial	Battery
Analog quartz	Quartz crystal	Integrated circuit divider	Dial chain (gears)	Hands, dial	Battery
Digital electronic	Quartz crystal	Integrated circuit divider	Integrated circuit divider	Digital display	Battery

an analog quartz watch, and the digital electronic watch. This comparison is shown in Table 1-1.

In the mechanical watch, the oscillating element is, of course, the balance wheel, as driven by the pallet fork. The frequency at which the balance wheel oscillates is divided down to one revolution per minute as required by the second hand. This is done by the timing chain, a set of gears which also transmits spring power to the balance wheel. In the Accutron® watch, the oscillating element is the tuning fork, as driven by a transistor.This frequency is divided by a ratchet wheel followed by necessary gears required to achieve one revolution per minute, again as required by the second hand. In the analog quartz and the digital electronic watch, the quartz oscillates at a much higher frequency than in the mechanical or tuning fork watches. This is normally at 32,786 vibrations per second (Hz).* But it can be at much higher fre-

*In the discussion on digital electronic watch technology, a few engineering terms will be used. *Frequency*—the power coming into your home and office in the United States is alternating current that goes through its cycle at the rate of 60 cycles per second. In 1960, an international conference on weights and measures made hertz the unit for cycles per second. It was named after Heinrich Rudolph Hertz, a German physicist who first detected the transmission of radio waves through space. The proper way to refer to your 60 cycle per second line frequency is now 60 hertz (Hz). For radio work, the electric energy can reverse direction many thousands or many millions of times per second, and in common use is the kilohertz (kHz), which is 1000 cycles per second, or the megahertz (MHz), which is one million cycles per second.

quency than this, even up to the megahertz (1,000,000 hertz) frequency range. This frequency is then divided down to one pulse per second to drive a stepping motor in the analog quartz or to supply other parts of the IC in the digital.

In the ordinary mechanical watch, the dial chain provides the proper gear relationship between the second hand, the minute hand, hour hand, and day/date. This same function is provided with essentially the same components in both a tuning fork and analog quartz watch. However, in the digital electronic watch, the relationships of seconds to minutes, minutes to hours, hours to date and day of the week, must all be done electronically. Normally, all of the information available within a digital electronic watch is not read out at the same time. There must be switching between functions to show, for example, hours and minutes with the press of one button, and month and date with the press of a second button. Seconds can also be called for in some comparable manner.

The power to operate the normal mechanical watch comes from the mainspring. For all the other watches, the power source is the battery. One battery is required for the tuning fork, the analog quartz, and the liquid crystal display digital watches. As will be explained later, the display voltage requirements of the LED watch generally require that two batteries be used.

The display of hours, minutes and seconds in both the mechanical and tuning fork watch is by means of the hands. Indication of day and date normally comes from a revolving disc behind the dial showing the

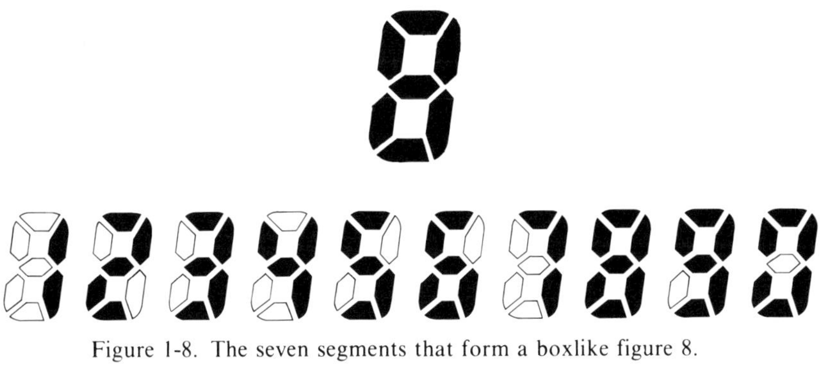

Figure 1-8. The seven segments that form a boxlike figure 8.

day and date. This is seen through a window in the face of the watch. In the digital electronic watch, the display is completely different. Both the LED and the LCD form numbers with something called a seven-segment display. These are the seven segments that form a boxlike "figure 8," as shown in Figure 1-8. The segments that are turned on to form the letters "1" through "0" are also shown.

Each of these components, from the quartz timekeeping base through the display and battery technologies, will be discussed in subsequent chapters.

2

Quartz Oscillators for the Electronic Watch

Since prehistoric times, the sun has been used by man to control his life. The sundial, the first timekeeping instrument, showed solar time. Clocks were developed to keep pace with solar time and to provide the time when the sun was not visible. But, as clocks became more accurate, the discrepancy between solar time and that kept by clocks became obvious. The problem was due to the fact that the path of the earth is not a circular orbit around the sun, but an elliptical orbit.

Since solar time was inadequate for man's use, timekeeping machines were made to accurately split the day into 24 hours, and the hour into 60 minutes. The timekeeping machines that keep pace are mechanical or atomic resonators which vibrate or spin at a fixed rate. The more independent these are to changes in temperature, atmospheric pressure and other miscellaneous changes, the better the device will be at tracking the rotation of the earth and providing accurate time.

The most widely used timekeeping machine, the balance wheel, depends upon a large mechanical motion at a relatively low frequency for its timekeeping base. A much more limited motion at a higher frequency is that of the metal tuning fork as used in Accutron® watches. The quartz crystal used in electronic watches is, in reality, just another mechanical resonator with an even smaller vibrating motion and a higher frequency. The average digital watch quartz vibrates from 32,768 to 4,100,000 times per second. It is made in three or four configurations, one of which is very similar in appearance to a metallic tuning fork.

A GOOD TIMEKEEPING MACHINE

We have all heard the argument: "My $9.95 Timex keeps as good of time as his $200.00 quartz." We know that they can both be adjusted at one point in time to be equally accurate. Their ability to maintain accuracy for long periods of time and in different temperatures and environments is strictly determined by the type and the material of the mechanical resonator. Figure 2-1 is arbitrary, but it does approximately represent the capability of these various timekeeping machines to keep time.

Several factors all contribute to the "goodness" of the resonator. One is insensitivity to its environment. A resonator that changes frequency due to changes in temperature, atmospheric pressure, or position does not make a good timekeeping reference for watches (or other clocks). This is true with the mechanical watch balance wheels, and it is also true for quartz watches. Beyond sensitivity to environment, however, there is another factor which is more dependent upon the material itself than anything else, and that is the ability of the material to vibrate or oscillate with a minimum loss of energy. Ignoring the loss

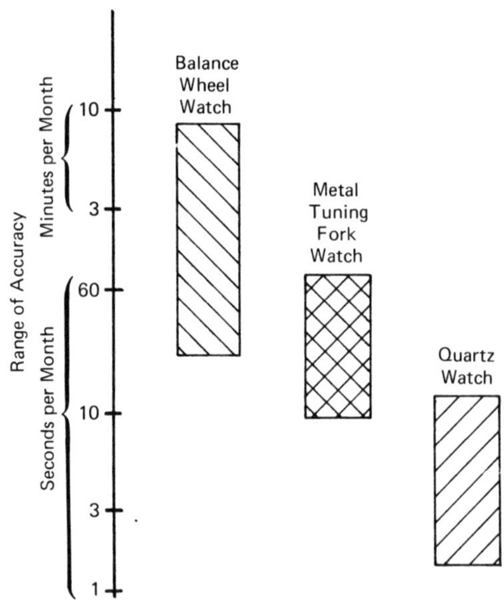

Figure 2-1. Relative accuracy of timekeeping.

of energy, a tuning fork made of lead could keep as accurate time as one made of the finest steel. However, the one made of steel will vibrate many thousands of cycles after being struck, and one made of lead would vibrate only a few. Having to supply extra energy to make up for losses is itself a major source of error in mechanical resonators, and therefore, the less external energy that is required, the more accurate the timekeeping base. The term expressing this inherent ability of materials to vibrate is referred to as "Mechanical Q." The lower the energy loss in each vibrational cycle, the higher the Mechanical Q. The Mechanical Q of metals is from a few thousand for ordinary metals to over 10,000 for the best tuning fork steel. Quartz at 32 kilohertz has a theoretical Q in vacuum of several hundred million, although the normal watch quartz has an operational Q of 100,000 or so.

There are materials that have a higher Mechanical Q than quartz. So why is quartz always used? Quartz is used because it can be made to be self-temperature compensated and it is electrically active (piezoelectricity will be explained in a later paragraph). When more accuracy is needed than is available from quartz, atomic frequency standards are the only available choice. There are complex, cumbersome machines that are not, even with the wildest imagination, suitable for use in electronic wristwatches. Every radio and television station uses quartz to determine its frequency of operation. Quartz crystals are excellent for frequency control and will probably not be replaced in this capacity in the foreseeable future.

Quartz is a piezoelectric crystal. That means that it has the ability to turn mechanical motion into voltage, and voltage into mechanical motion. Several other crystal substances besides quartz have this capability, but none of them possess the very high Mechanical Q and temperature compensation properties of quartz. In the tuning fork in the Accutron watch, two small coils are required to provide driving power to the tuning fork. In quartz, the driving power in and frequency output are all provided by the piezoelectric effect.

DIFFERENCES IN WATCH QUARTZ

Assuming a quartz crystal will be used for a watch, the first criterion will be the selection of quartz frequency. Quartz crystals are manufactured to operate from the audio frequency range (a few thousand hertz) to the upper end of the television band (150 megahertz). The selection of a frequency for watch applications is therefore arbitrary and depends upon factors such as minimizing the physical size and cost of the quartz. The impact of the quartz frequency on the accuracy of the watch is minimal. Of the two most accurate quartz watches that have been manufactured, one has the quartz oscillating at 16 kilohertz and the other uses quartz at above 4 megahertz. Accuracy is in no way related to the operating frequency of the quartz, but is strictly dependent upon the quality of engineering and care that goes into the use of quartz at any frequency.

The majority of quartz crystals being used in digital (and analog quartz) watches today operate at 32 kilohertz, and our discussion of quartz will be centered on quartz that operate at that frequency. Quartz that operates at 16 kilohertz is very similar to the 32 kilohertz quartz, except slightly larger. Quartz that operates in the 700 kilohertz and 4 megahertz range is distinctly different in operating mode and appearance.

From the physical appearance of most quartz crystals used in watches, it is obvious that they are just another mechanical resonator. Figure 2-2 shows a tuning fork configuration quartz that looks like a tuning fork. The other configuration popular with United States manufacturers is the flexural mode. This is a vibrating mode very similar to a xylophone bar. This configuration is shown in Figure 2-3.

A variation of the tuning fork quartz has appeared and found wide acceptance. This is a thin version of the quartz—referred to as a "leaf" quartz. Its advantage is that it can be processed on a "slice" of quartz much as integrated circuits are processed on a slice of silicon. Figure 2-4 shows such a slice prior to the separation into individual tuning forks. Figure 2-5 shows the mounting of the quartz in its package, and Figure 2-6 shows a packaged quartz. There is a glass window as a part of the quartz package so the exact frequency of operation of the quartz can be adjusted by use of a laser beam after it has been sealed in the package.

There is the obvious potential of lower cost with the leaf quartz. On the negative side, this quartz generally has a narrower tuning range and is more sensitive to package leaks than quartz tuning forks sawed from a quartz ingot.

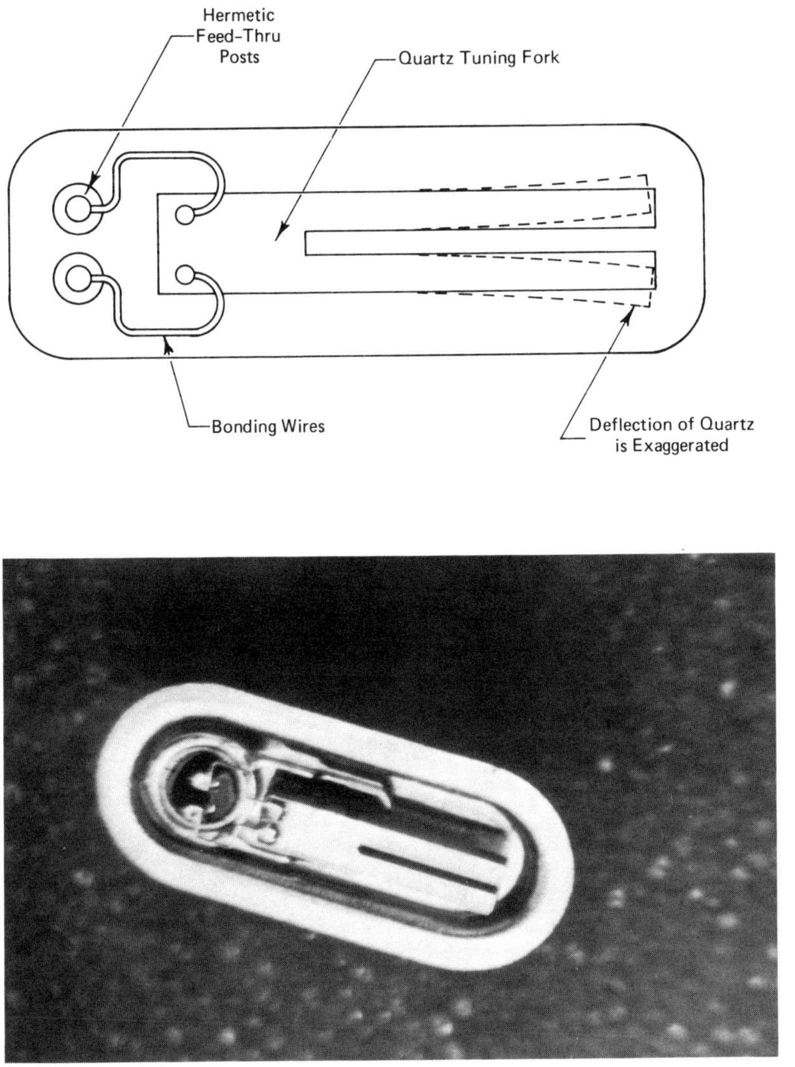

Figure 2-2. Schematic drawing and photograph of tuning fork quartz. (*Courtesy* of Matsushima Kogyo Co., Ltd.)

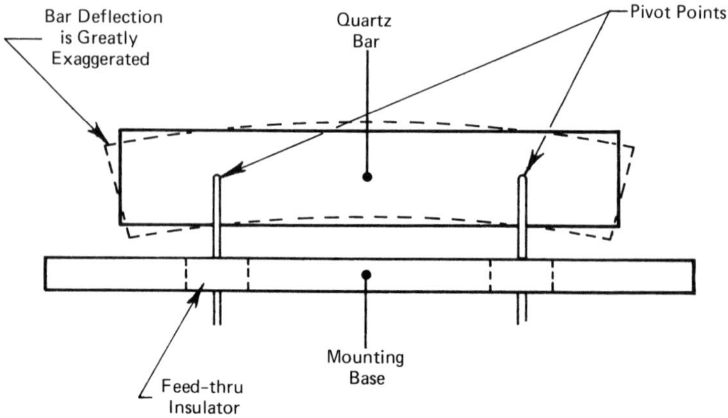

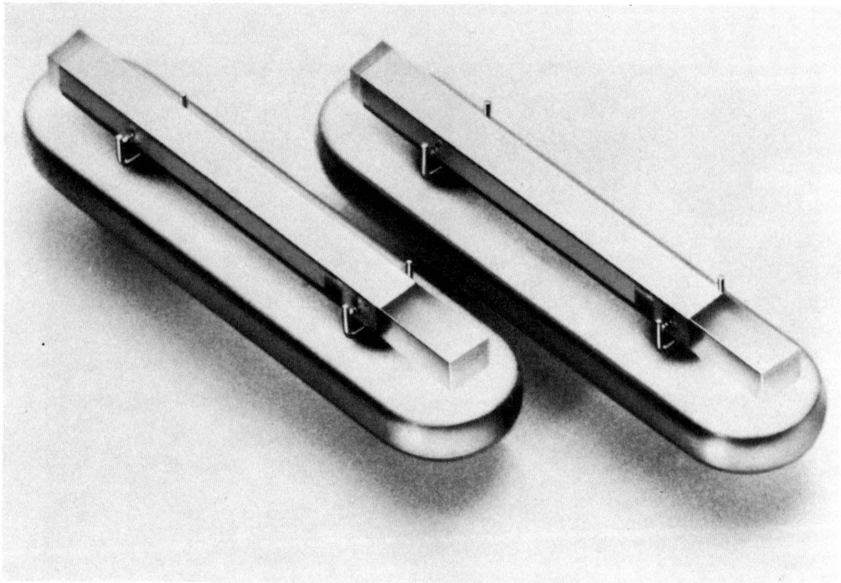

Figure 2-3. Schematic drawing and photograph of flexural mode quartz. (*Courtesy* of Omega, SSIH-Quartz Division.)

Figure 2-4. Leaf quartz slice prior to separation. (*Courtesy* of Statek Corporation.)

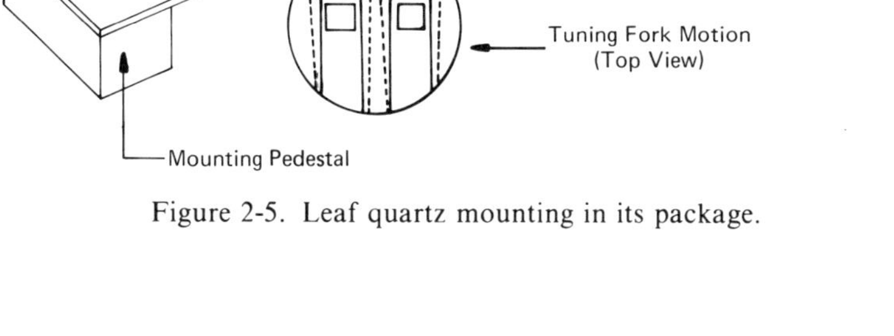

Figure 2-5. Leaf quartz mounting in its package.

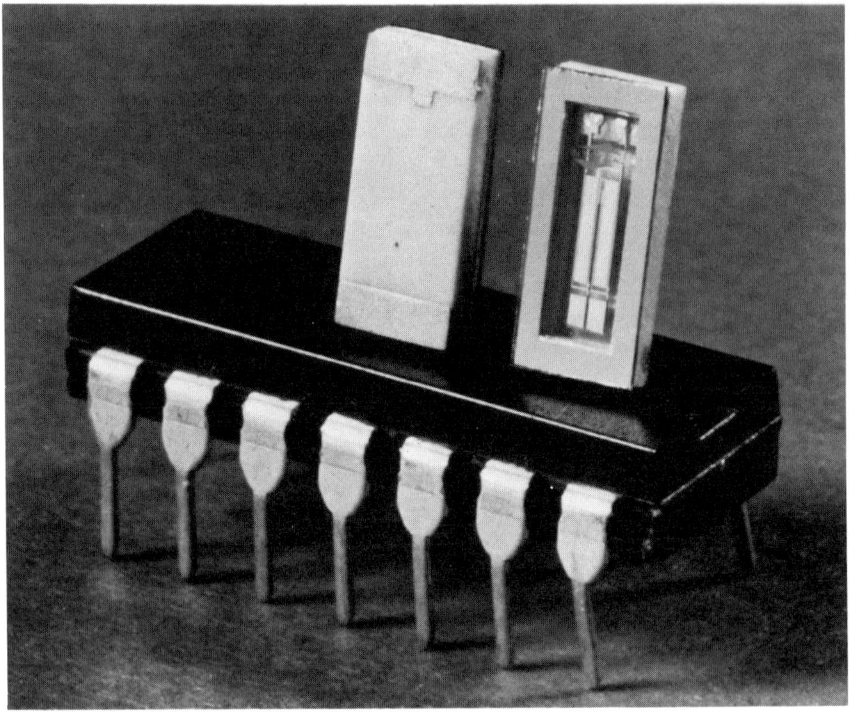

Figure 2-6. Packaged quartz. (*Courtesy* of Statek Corporation.)

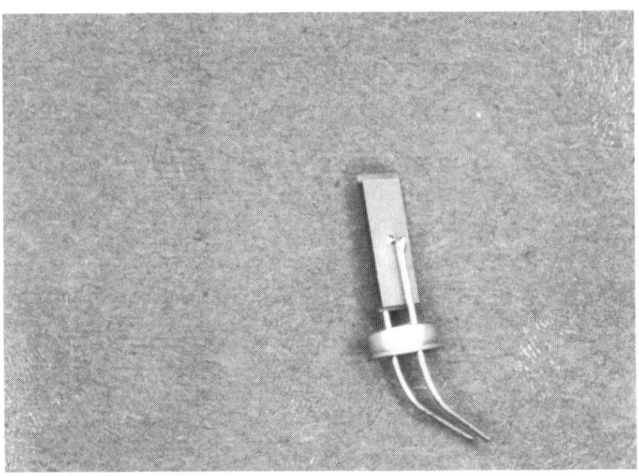

Figure 2-7. Hughes Aircraft Company quartz crystal.

A quartz crystal manufactured by Hughes Aircraft Company for operation at 786,432 hertz is shown in Figure 2-7. This frequency is in the AM broadcast band, and over twenty times the frequency used on most competing watches. The advantage of this type of quartz is greater stability with temperature. The disadvantage is the higher current required by the integrated circuit due to the high frequency of oscillation.

CHARACTERISTICS OF QUARTZ IN A WATCH

Except as discussed, the characteristics of a quartz used in a watch are independent of the configuration of the quartz. These shared characteristics determine the operational performance of quartz in a wristwatch. One shared characteristic of all quartz is aging, something that occurs during the entire life of the watch. If a quartz watch changes by one second per month due to aging during the first month of operation, then it will also change by one second per month during the following two months, and one second per month the next four months, etc. This logarithmic type of change is shown in Figure 2-8, and it is normally a change from one second per month to four seconds per month in timekeeping per decade of change in time. By the time a quartz watch is one year old, the error due to aging problems is negligible.

The temperature characteristic of quartz varies from type to type as does its size. There are vibrational modes possible in quartz that vary less than one second per month over a temperature range of 0 to 50° C. Quartz having this broad a temperature range is much larger than other types of quartz for a given frequency. The temperature controlling effect of wearing a watch on your arm allows a smaller quartz to be used but one that also has a narrower temperature range. Typical temperature characteristics are shown in Figure 2-9. This indicates that the watch would lose one second per day from its nominal timekeeping accuracy when the temperature is reduced to 10° C and also would lose one second per day when the temperature is raised to approximately 43° C. Operation of the watch on the wrist tends to maintain a reasonably constant temperature even though the wearer may be in the arctic or in a tropical jungle. However, it the watch is removed either where it is hot or cold, the timekeeping ability is affected. A small number of

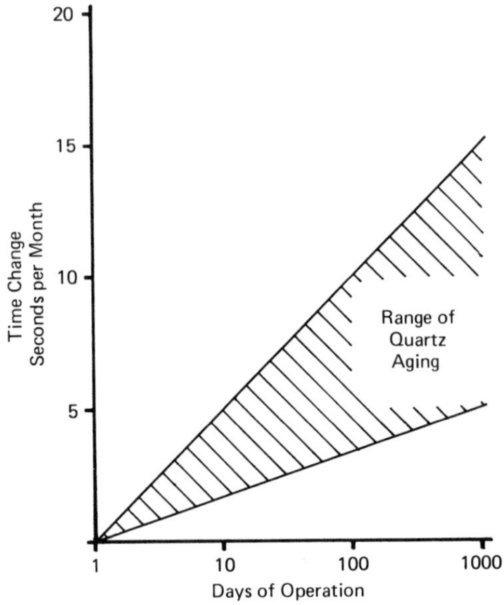

Figure 2-8. Aging effects of quartz.

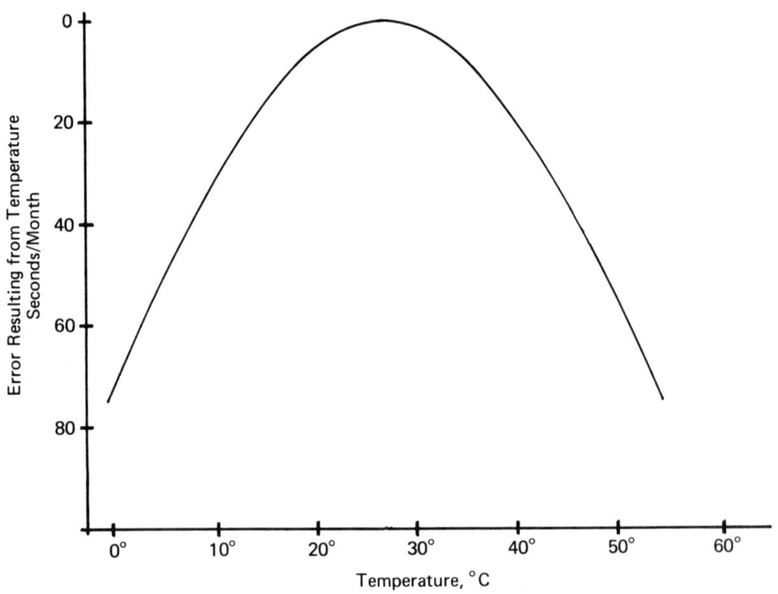

Figure 2-9. Typical quartz temperature characteristics.

expensive quartz watches have either temperature compensated quartz oscillators or higher frequency quartz that has less temperature variation. These provide ultimate accuracy across a broad temperature range.

Dropping a quartz watch causes changes in timekeeping accuracy just as it does in watches having a balance wheel. There are two kinds of damage that can occur at a given shock level. In one type, the quartz is disturbed but not permanently damaged, as shown in the top line of Figure 2-10. The other type damage results in a permanent change of the quartz and it will not recover to its original timekeeping accuracy without the trimmer being readjusted. The curves are shown at a 4000 g shock, somewhat more severe than a 3-foot drop to a hardwood floor, as specified in the FTC standards.

Quartz does not suffer from significant gravitational effects but it is sensitive to the ambient pressure. For this reason quartz is mounted in a hermetically sealed can, either under a complete vacuum or under a partial vacuum with helium gas in the can. One of the failure modes of quartz is for a slow leak of air into the can to occur. This will cause a 20 to 30 second per day loss by a watch. The curve of pressure versus timekeeping for the quartz is shown in Figure 2-11.

The effects discussed in Figures 2-7 through 2-11 are primarily those

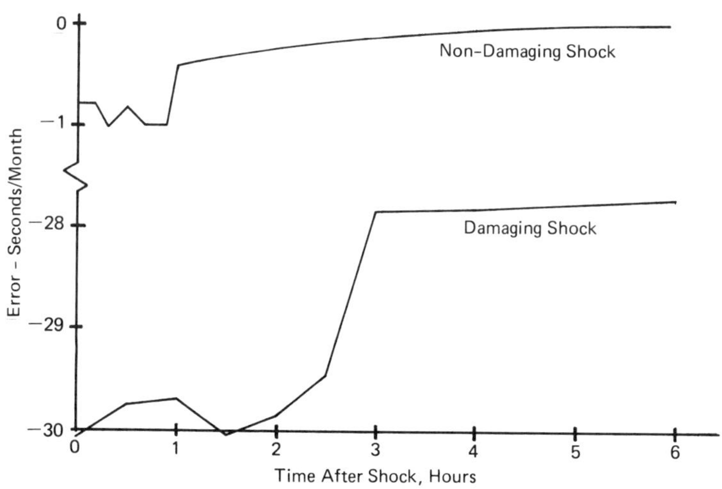

Figure 2-10. Effects of shock on frequency.

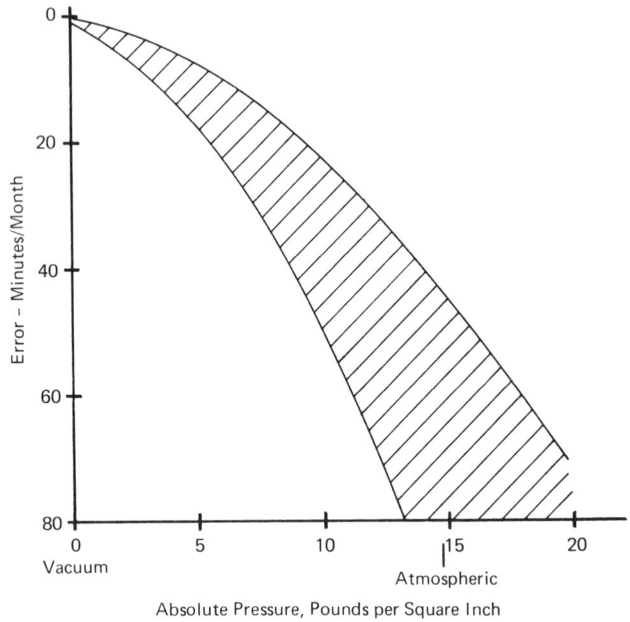

Figure 2-11. Effect of internal pressure on quartz frequency.

of the quartz and are reasonably independent of the circuitry in which the quartz is operating. The factors within the digital electronic watch which also affect quartz performance are summarized in Figures 2-9 through 2-11.

THE TRIMMER CAPACITOR

The dominant factor, other than the quartz, in determining the frequency at which the watch operates is the trimmer capacitor. The trimmer capacitor is very similar in electrical performance to the tuning capacitor used on AM and FM radios. Since it must be much smaller than the device used in radios a different construction is required. Figure 2-12 shows a trimmer capacitor. The "half moon" metal pattern on the top is movable as the trimmer is rotated with the screwdriver. There is a half moon metalization on the bottom. When the top and bottom are aligned, the watch is at the slowest setting; when the top

Figure 2-12. Trimmer capacitor.

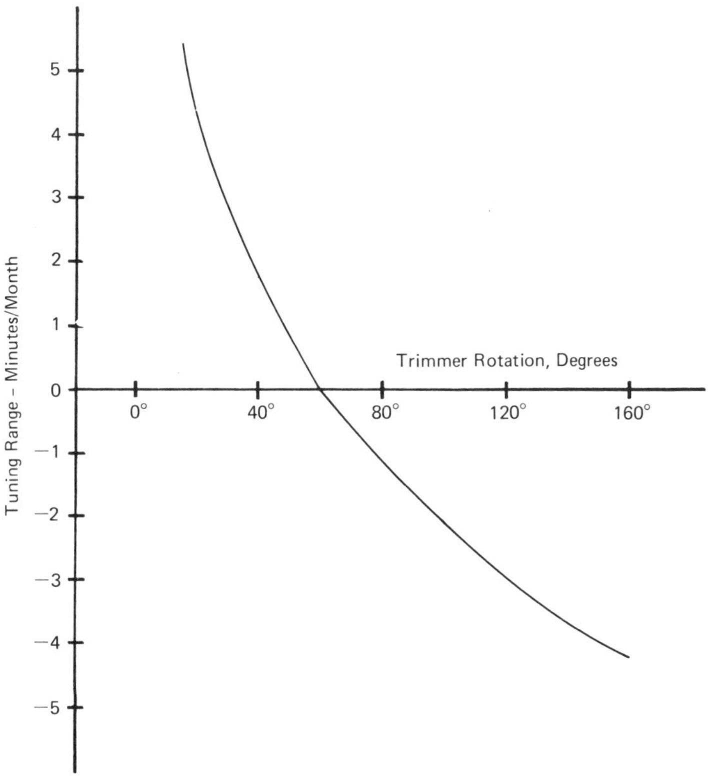

Figure 2-13. Tuning of a quartz crystal.

is rotated so the metal portions do not overlap, then the watch is at its fastest.

One problem with the trimmer is that if the watch is at the slowest setting, turning either clockwise or counter-clockwise will result in faster settings. The mechanical watch is adjusted faster or slower with the regulator. It has an indicator which shows the direction it must be moved to make the watch beat faster. With the present trimmer capacitors the only way to determine the direction for a faster setting, for example, is to use a quartz watch tester (to be discussed in Chapter 9) and observe the results when the trimmer is adjusted. A trimmer with indication of Fast and Slow has a place in the future of digital electronic watches.

Timekeeping of a watch can normally be varied plus or minus 10 seconds per day by use of the trimmer capacitor. A tuning range this broad is required in order to compensate for the manufacturing tolerance of the quartz and provide correction for all other effects listed in this chapter. A normal curve of timekeeping versus trimmer capacitance is shown in Figure 2-13. Notice that the change in

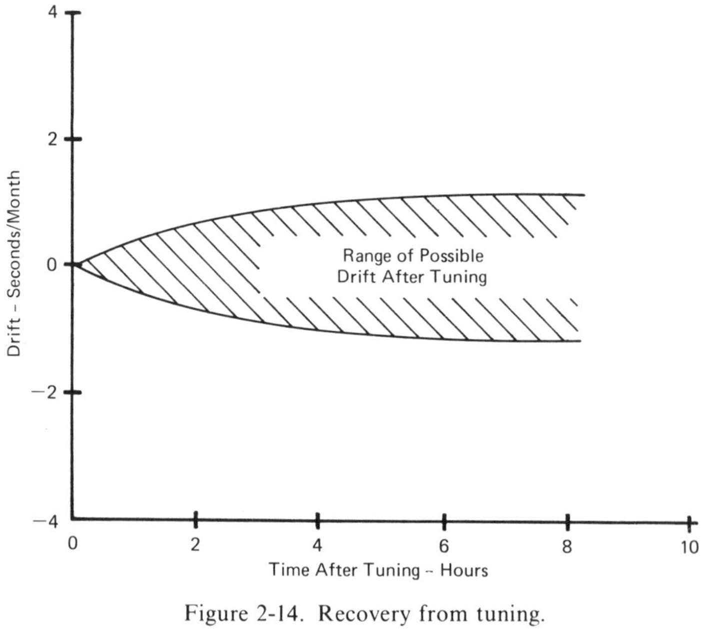

Figure 2-14. Recovery from tuning.

frequency is much greater at the low capacitance end than it is at the high capacitance end, and the effect for the watchmaker will be that in a certain portion of the trimmer capacitor tuning range, tuning to exact frequency will be very difficult, and in other portions it will be significantly easier.

As was evident in Figure 2-10 on the effect of shock and recovery from shock, there is also a recovery from changing the frequency of the timmer. This effect as shown in Figure 2-14 is much smaller. To adjust a watch to accurate timekeeping requires setting the watch, letting these effects age out, and then making the final precise adjustment to operating frequency.

THE EFFECT OF BATTERY ON ACCURACY

Another source of error that is dependent upon both the battery and the oscillator portion of the integrated circuit is the effect of battery voltage upon the timekeeping ability of the watch. Figure 2-15 shows how the battery voltage varies throughout its life. Figure 2-16 shows

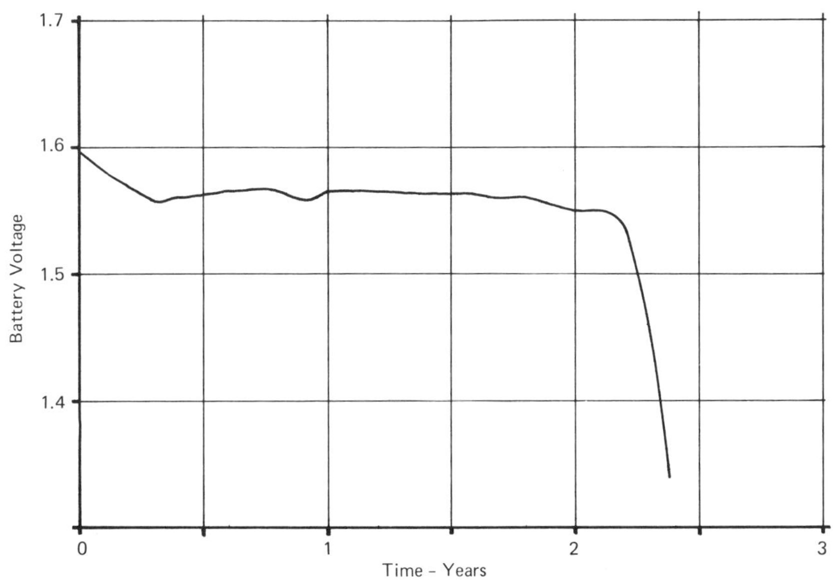

Figure 2-15. Battery voltage variation.

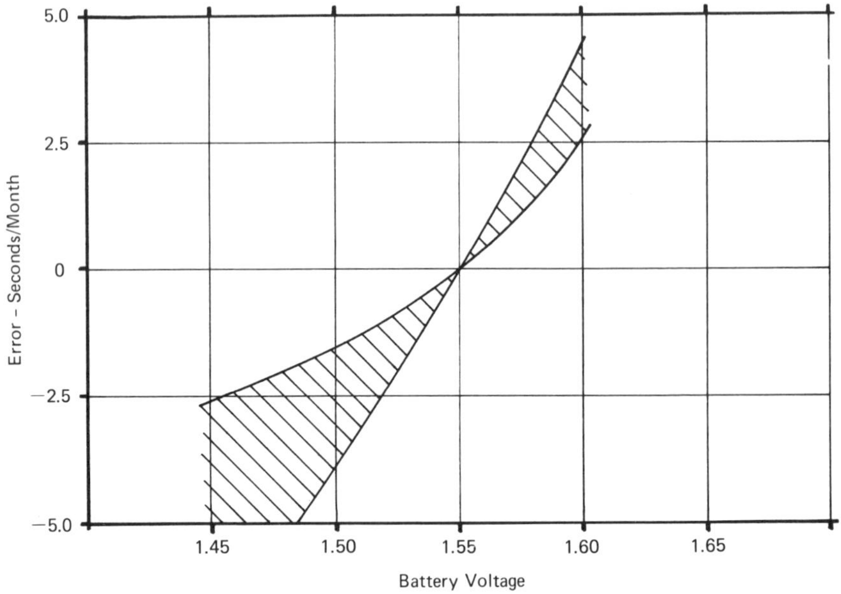

Figure 2-16. Accuracy as influenced by battery voltage.

the effect of battery voltage variation on timekeeping. It is necessary that before a precise adjustment of timekeeping is made, that the battery condition be ascertained and that a new battery be used if there is any doubt at all about the life remaining in the battery.

HOW ACCURATE WILL IT BE?

The foregoing discussion is centered on accuracy. Can the watch be adjusted to accurate time? Will it stay on accurate time? The key questions that should be asked are as follows.

The Quartz	Is it properly designed?
	Is it manufactured and tested with the proper controls?
	Has it been aged (operated 3 to 6 months)?
The Electrical Circuit	Is it properly designed?
	Is the trimmer capable of the necessary resolution for adjustment?

Is the oscillator circuit temperature compensated (to compensate for quartz variations)?

The Module Design

Does the module reduce the shocks seen by the quartz?

Is the voltage seen by the oscillator voltage regulated?

Are key portions of the oscillators circuit properly shielded from stray interference?

A yes answer to all of the aforementioned questions occurs only with a few watches—it is simply too expensive for everyday watches. Below a correlation is made between the attention given to design factors and the accuracy of a watch.

The best designs where cost is no factor
Accuracy ±1–2 seconds/month.

Good design in competitive price watch
Accuracy ±10 seconds/month.

Complete lack of attention to the above factors
Accuracy ±30 seconds/month.

Even with quartz, what you get depends upon how hard you try.

3 The Integrated Circuit

THE WATCH CIRCUIT

When looking at an electronic watch module, the integrated circuit is an obscure component that does a job, which in many ways, is as equally obscure. In this chapter, the jobs, the mounting, and the connection of an IC to other watch components, as well as some of the failures of an IC will be examined.

The integrated circuit does all the functions that are performed in the normal watch by the miscellaneous gears from the balance wheel to the hands, including the various setting mechanisms. It is the integrated circuit that makes the quartz oscillate and divides the quartz frequency to one pulse per second. In the analog quartz watch this is all the integrated circuit does. In the digital quartz, however, the IC further counts seconds, minutes, hours; contains all of the setting functions and provides the voltages and currents required to drive the LCD (liquid crystal display) or the LED (light emitting diode) display. Only since 1975 has it been possible to perform all of these functions in one integrated circuit.

Going back in time, to do all the logic and counting functions that a watch integrated circuit does using vacuum tubes would have required between 500 and 800 vacuum tubes (it would fill two cabinets six feet tall), and would require about 10,000 watts (W) of electric power or about the power consumption of 10 electric irons. Even doing a watch circuit function with normal integrated circuits, the kind that military equipment and computers are made of, would require 40 to 50 ICs and a power drain of maybe 15 watts. While this is a great improvement over vacuum tubes, the space required and the amount of power required are not consistent with the job of an electronic wristwatch.

What size and power consumption is consistent with the job of an electronic wristwatch? For this, we refer back to the mechanical wristwatches as discussed in Chapter 1. Larger diameter watches have internal volumes of 6 to 10 cubic centimeters. For a watch using two of the largest batteries of suitable size (these store 225 milliampere-hours each), there would only be 450 milliampere-hours available.* Since there are 8760 hours per year, for these batteries to last one year, there can be no more than 0.051 milliampere (51 microampere) current drain at 1.5 volts (V) for the watch.

Using two batteries of this size results in a thick watch at the very large end of the scale. It is more reasonable to expect the use of batteries with capacity totaling about 100 milliampere-hours in a man's watch, and 30 to 40 milliampere-hours in a ladies watch. The total power available to drive the average watch must, therefore, be less than 12 microamperes (μA) for the battery to last a year, and for a ladies, less than 4 microamperes.

To do this requires a factor of about two million improvement in current consumption compared to the integrated circuits that were available ten years ago. These are now available.

A breakthrough of this magnitude generally results from a completely new way of doing things. For low power integrated circuits for watch application, there has been developed not just one approach to doing circuits at this low power, but two completely different approaches—each with its own advantages and disadvantages.

A LOOK AT IC TECHNOLOGY

In order to discuss integrated circuits, it is necessary to go briefly into some of the "alphabet soup" terminology. The following list should be used as a brief glossary of terms for understanding the IC:

*A milliampere (mA) is 1/1000 of an ampere of electrical current. For comparison, a penlight bulb requires about 30 milliampere current. A milliampere-hour (mA h) is a unit of energy for a battery. This would be one milliampere for one hour. One penlight AA cell contains about 600 milliampere-hour of energy and will power a 30 milliampere bulb for about 20 hours.

1. The first transistors in commercial production were *bipolar* transistors. There are two types of bipolar transistors: NPN for normal operations with a positive power supply, and PNP for use with negative voltage power supply.
2. The first integrated circuits in large volume production were bipolar. These were generally manufactured using NPN transistors, and would operate off a single 5 volt power supply for logic circuits.
3. About the time of the development of the integrated circuit, there was the development of the *Field Effect Transistor* (FET). These are devices that depend upon electric fields in the semiconductor to turn on and off the current flow, rather than using a small current flow to control a larger current flow as is the case in bipolar transistors and ICs. FETs were made both in type N material (which had the anode positive), and type P material (which had the anode negative), similar to bipolar transistors.
4. Between 1965 and 1970, manufacturers began to build and sell field effect integrated circuits. These were generally of the type having a metal control element, a silicon oxide layer between the control layer and the silicon underneath. This stack gave rise to the word, Metal-Oxide-Semiconductor, or MOS integrated circuit.
5. Early ICs, mainly for calculators and other computing applications, were type P-MOS structures, or P-MOS.

The listing of the various IC technologies needs to be interrupted briefly here by a discussion of logic. Logic is performed in semiconductors (and vacuum tubes, too) by having the device turned "on" for one logic state and turned "off" for the other logic state. Two of these logic elements are often connected in pairs so that when one is turned on, the other one is always turned off—a configuration called a "flip-flop." This is the type of element used extensively in watch circuits to divide the frequency from the quartz oscillator down to the seconds, minutes, hours and months, as required for the watch operation. This approach to logic means that for every transistor that is turned off and is not drawing current, there is one transistor turned on and drawing current. As a result, P-MOS IC performing a watch function would require approximately 1/5 watt for the logic functions.

(From two batteries, there is a maximum of 76 millionth of a watt available continuously for one year operation).

With increasing understanding of semiconductor processes, it became possible to manufacture type N-MOS devices and type P-MOS devices on the same IC bar. A logic element, instead of the choice of being on or off, could therefore be made by placing a type P-MOS device in series with a type N-MOS device, as shown in Figure 3-1. When one device was turned on, the other was turned off; therefore, current never flowed. The only current required for this device is switching current.

6. The term expressing the use of two MOS transistors of different polarity in series is complimentary MOS, or C-MOS. A C-MOS integrated circuit at rest will draw only a few billionths of a watt power for its leakage, and when operating, as driven by a 32 kilohertz quartz for watch application, may draw only 5 to 10 millionth of a watt.
7. In the meantime, bipolar integrated circuit terminology also progressed, and it became possible to operate bipolar logic elements where, even though a device needed to be drawing current, the current that it drew was only a few billionths of an ampere. A way of manufacturing bipolar circuits was also developed that allowed as close a spacing in bipolar as was

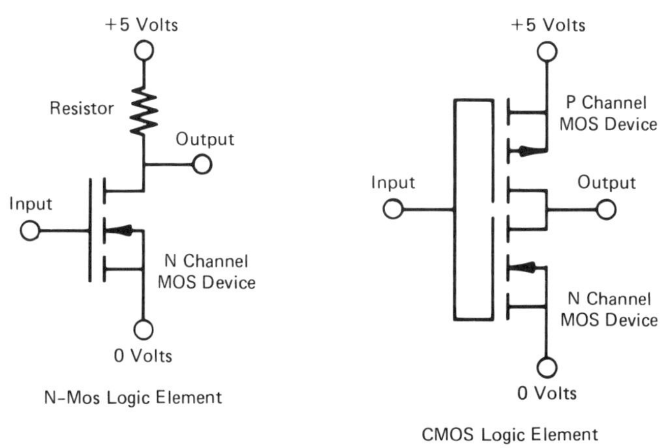

Figure 3-1. Logic element in MOS and C-MOS.

possible in MOS. This bipolar circuit configuration has also been used for watches, and is known as *Integrated Injection Logic*, or "I^2L."

WHICH IC FOR THE WATCH

No one device is a panacea for all problems. The C-MOS represents the ultimate in low-current operation, but its ability to supply the high currents required for LED displays was inferior to that of bipolar ICs. Many ICs for LED watches required external transistors to drive the LED display. The I^2L integrated circuit was capable of currents necessary for driving the LED, but could not achieve as low a power consumption as was achieved by the C-MOS when driving a liquid crystal display.

At present, the status is that both C-MOS and I^2L are being used for LED digital watches. There is no advantage to the customer or the watchmaker of one over the other, it strictly represents an economic trade-off on the part of the manufacturer. Accuracy, reliability, and other consumer and service factors should be equivalent. The slightly higher power consumed by I^2L is usually not important in the LED watch due to the power consumed by the display.

In the liquid crystal watch, C-MOS is used exclusively. It is not expected that the I^2L circuits will be developed for liquid crystal watches.

We have already compared the functions performed by an integrated circuit to those performed in a mechanical watch. A diagram of the functions performed in an electronic watch IC is shown in Figure 3-2. The oscillator portion of the circuit is very similar to the oscillator circuitry used to drive an Accutron-type tuning fork. The countdown from 32,687 kHz is in 15 stages of divide-by-two to reach one pulse per second. If we remember that the integrated circuit can only be divided by two, then some of the constraints become obvious. A divide-by-ten, for seconds, is 4 stages of divide-by-two (which gives a divide-by-sixteen), but with the appropriate interconnections to reset to 0 when the count reaches 10. The 10s for seconds is three stages of divide-by-two with connection to reset to 0 when the count reaches 6. Counting for minutes, hours, and date goes along very similarly. The only other portion of the integrated circuit which does not have a

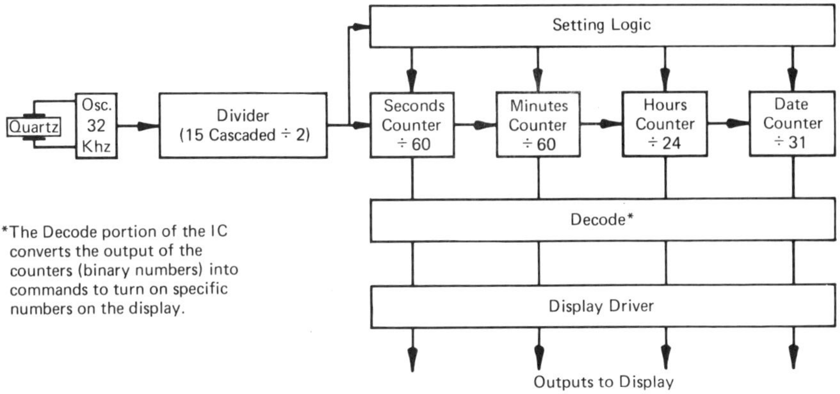

Figure 3-2. Block diagram of an IC.

direct counterpart in the mechanical part is the voltage multiplier that is required to power the liquid crystal display.

A divide-by-two stage of the integrated circuit is shown in Figure 3-3, and the entire watch integrated circuit is shown in figure 3-4.

Once we have the integrated circuit, then what? In an examination of watch modules, nothing that looks like an integrated circuit is visible. This is because the integrated circuit must be mechanically protected so that handling in manufacture does not scratch or smear the metal pattern on the integrated circuit or pull off the wires that make contact with the IC. It must also be protected from atmospheric

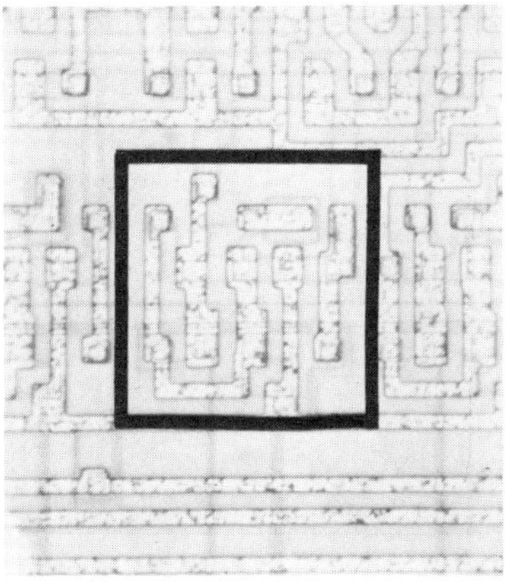

Figure 3-3. Microphoto of divide-by-two portion of watch IC.

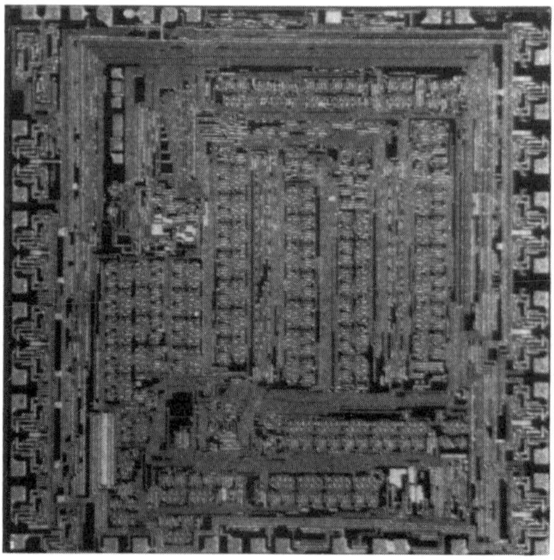

Figure 3-4. Microphoto of watch IC.

Figure 3-5. Watch IC mounted in hermetically sealed package.

problems. Pure distilled water condensing on the IC is not too much of a problem, but the water that leaks into a watch is not pure. It often contains some salt and repeated condensation and drying of water containing salt on the IC can provide unwanted conductive paths between the metal conductors on the IC, and can also actually cause a failure within the silicon itself over a longer period of time.

The first protection used for integrated circuits was a sealed package, where contact was made between the package leads and the integrated circuit with fine gold wire. It was filled with an inert gas and then sealed to be airtight. During the last 10 years, plastic packaging of integrated circuits (where the plastic is molded to make a solid block with the IC) has become very popular and the reliability data on integrated circuits completely surrounded by plastic has been as good as the reliability of the hermetically sealed ICs. Both types of construction are used in electronic wristwatches. A watch module using a packaged IC is shown in Figure 3-5, and a watch module having a plastic encapsulated IC is shown in Figure 3-6.

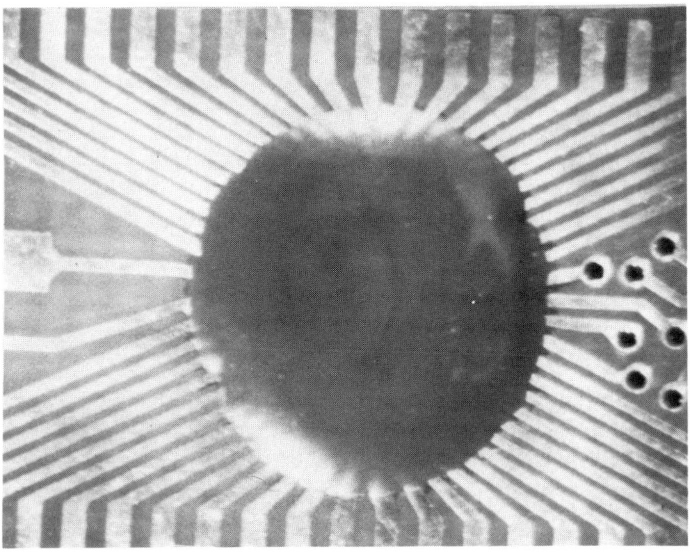

Figure 3-6. Watch IC with open encapsulation.

MAKING THE DIGITAL WATCH SERVE THE CUSTOMER

Even with quartz, the IC and the display present, it must be "tamed" to be a watch usable by the customer. The taming is centered in two areas of endeavor; deciding what functions to put (and not to put) into a digital electronic watch, and deciding what is the best approach for both the customer and the manufacturer for setting purposes.

The trend in business has always been that a business starts simple, rapidly gets more complex, and then a point of diminishing sales appears or diminishing return on the investment is reached. In its short life time the digital watch already has been through several generations. Going back in history, the first digital watches displayed hours and minutes only. The next step was the addition of seconds, primarily in the LED, followed closely behind by a 31-day-month calendar, as is available in mechanical watches.

While it is difficult to make a calendar in mechanical watches that compensates for the number of days in the month, it is fairly easy in electronic watches. The five function watch, which shows month and date, enables the use of something that is referred to as a "smart calendar." It knows that not all months have 31 days and can put in the appropriate 28 days for February and 30 days for those remaining months. The smart calendar has been available in both the LED and the LCD since mid-1975.

Adding the day of the week was more of a display problem than it was an IC problem. There have been two schools of thought on the day of the week. One is to have a symbol or contraction representing each day of the week manufactured in the display and then to either illuminate the day symbol or give some other indication. An example of this type of watch is shown in Figure 3-7. A square is drawn around the appropriate day. One advantage is that the language can be changed by simply changing the printing on the display. No IC or connection change is needed.

Another approach to the day of the week (as well as the name of the month) is referred to as alphanumeric. The display is modified so that it can form both alphabet letters and numbers. The same display that shows the time can show the day of the week and date when the function button is depressed. An example of this is shown in Figure 3-8. The main drawback to this approach is increased complexity of the display, and every language requires a different IC modification.

Figure 3-7. Watch showing day of week by "Flag." (*Courtesy* of Seiko Time Corporation.)

The digital watch is an accurate timer; therefore, adding stop watch functions to the digital watch to produce a chronograph is comparatively straightforward. Timing to the 10ths and 100ths of seconds can be accommodated by changes in the IC, and keeping track of elapsed time, as well as the local time, requires the addition of seconds and minute counters and the ability to display the information in these counters upon demand.

The situation is very similar on the alarm watch. An alarm watch requires only the addition of a counter into which the alarm time is set. The information stored in these counters is continuously compared with the time and when the two match, the integrated circuit produces

Figure 3-8. Alphanumeric display of day/date on a liquid crystal watch.

a signal, which is used to drive a buzzer. The main problem involved in the alarm watch has been in the electromechanical portion, the buzzer.

To go beyond the functions that are available in mechanical watches is a natural path for the digital watch, but the needs of the consumer are not yet known and much experimentation is required. This is especially true with watches having broad multiple function capabilities. Several manufacturers produce watches with broad, multifunction capability. These, besides showing local time, can function as a counter, as a stop watch, as an alarm watch, as a calendar, and can show the time in different zones. Such a watch is shown in Figure 3-9. Long-term appeal of these watches will depend upon their usefulness to the user. This, in turn, depends upon both the capability of the watch and the user's ability to operate it. The user will not carry an instruction manual for his watch.

There has been a recent development that makes "customizing" the functions of a digital watch much easier. This development is the microprocessor-based integrated circuit. Watch circuits have been, until this time, specially designed for each application. To introduce a different watch would require the development of a new integrated

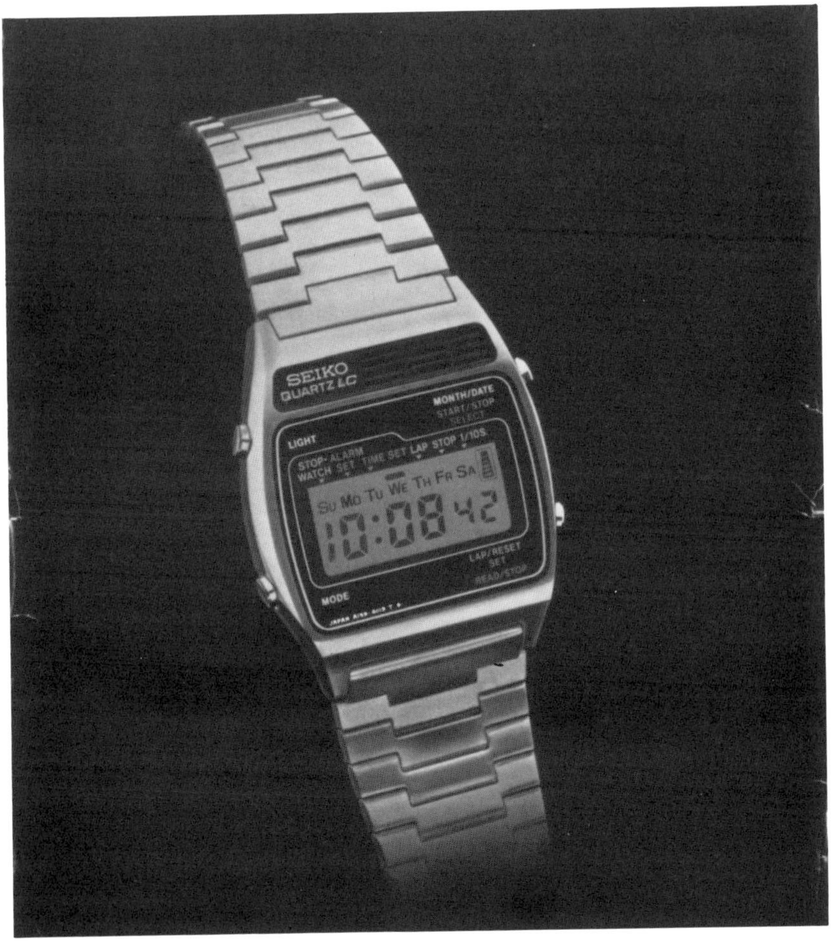

Figure 3-9. Multifunction watch. (*Courtesy* of Seiko Time Corporation.)

circuit. Each new IC required many months and many thousands of dollars.

With the recent advent of electronic calculators came a new way to make circuits perform functions. It was a way the circuit could be instructed (by use of information contained in a memory on the integrated circuit) as to what job to do. This same approach has been applied to watch integrated circuits. One such circuit is the Intel 5830 family of circuits. This integrated circuit is shown in Figure 3-10.

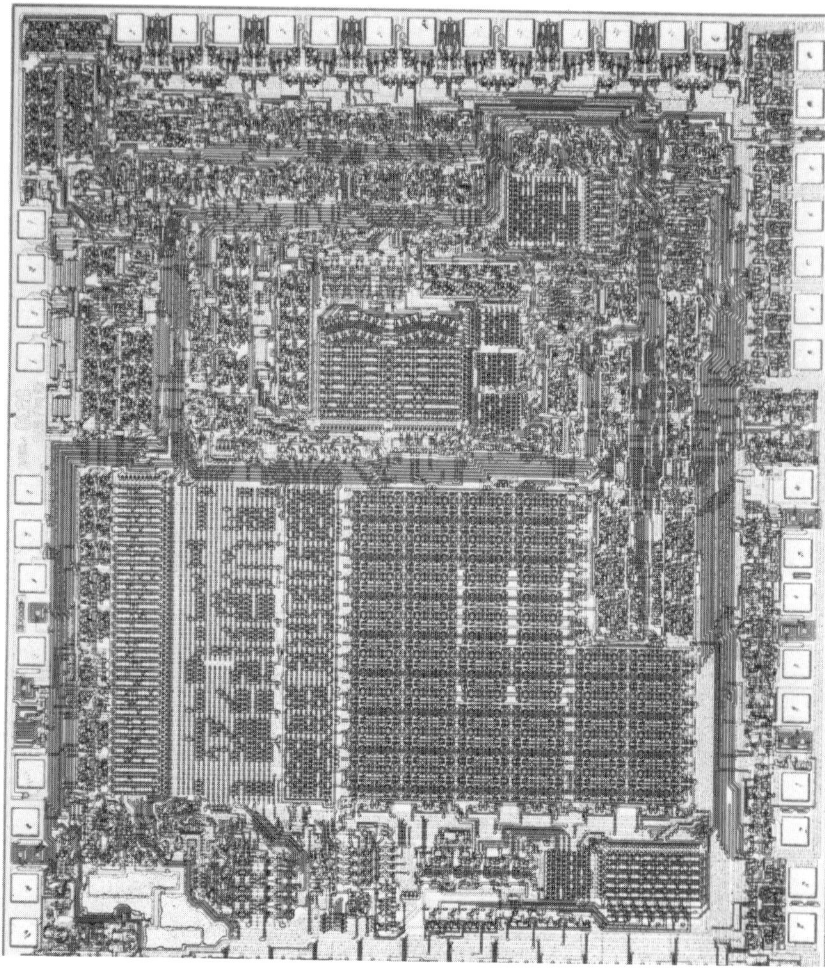

Figure 3-10. Microprocessor IC chip. (*Courtesy* of Intel Corporation.)

Comparison with the photo of a conventional IC as shown in Figure 3-5 will show a completely different look. The major portions of the microprocessor IC are shown in Figure 3-11. This shows the display driver across the top and bottom of the IC, a memory section (Random Access Memory or RAM) in the upper counter part of the IC that remembers what the last second, minute, hour and date was, and stores that information until it is ordered changed by other parts of the integrated circuit.

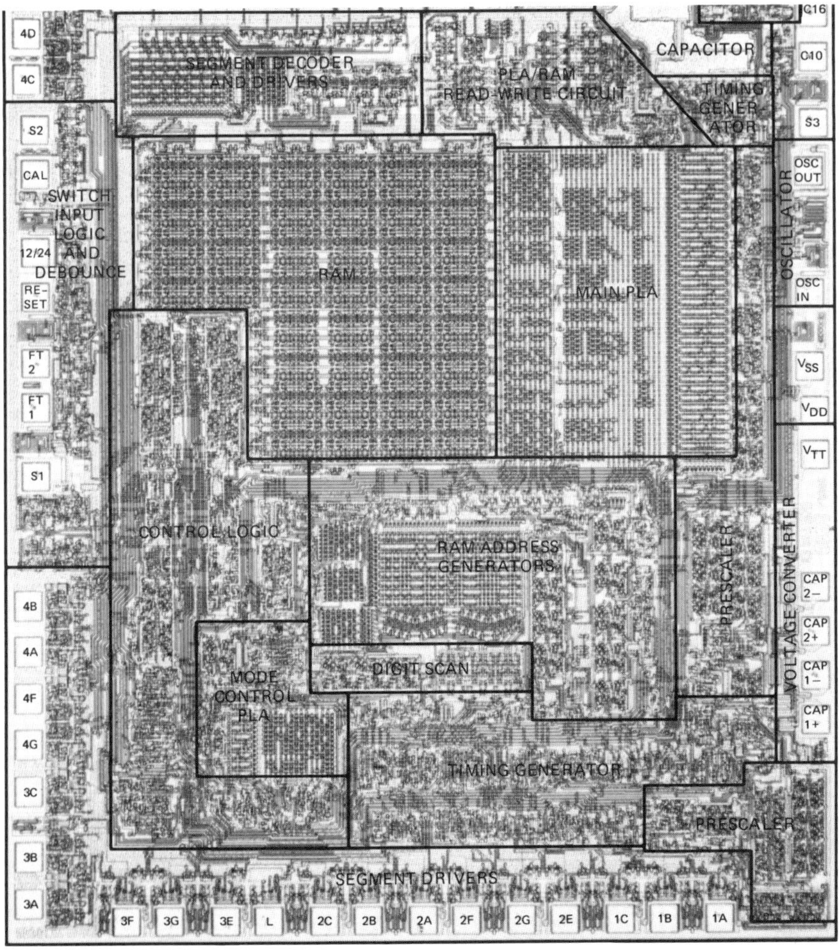

Figure 3-11. Major portions of the microprocessor chip. (*Courtesy* of Intel Corporation.)

Next to the memory is a Programmable Logic Array (PLA) which contains the programming that tells the IC which function to perform and which outputs are used for what purposes."

This type of integrated circuit requires more silicon area to perform a fixed job than is required by the type of circuit shown in Figure 3-5; therefore, the cost will be higher. For short production runs of special functions and customizing the appearance, the programmable IC has a permanent place in the future of digital watches.

SETTING THE DIGITAL WATCH

Accompanying the changes in watch function, there have been changes in the way that the watch is set. The first hour/minutes only digital watch used two push buttons. One to advance hours, the other for minutes, and pushing both buttons simultaneously set seconds to zero and held them there until the buttons were released. An early LCD that showed seconds, minutes, hour and date had four push buttons on the back of the watch. The push buttons for date, hours and minutes advanced at one count per second when the push button was depressed, and the push button for seconds reset the seconds to zero and held them there as long as the push button was depressed.

With the coming of the function button to call up different functions of the watch, a sequencing type of setting became the practical way. The most popular means of setting now in use is the sequential, using a recessed corrector push button. One push of the corrector button causes the display to show only the month. Month correction is made by depressing the function button causing the month to change at one month per second. Again depressing the corrector button will cause date to be displayed, and then further pressing of the corrector button will call up hours and then minutes and then return to normal operation. Accurate time will not be lost unless the minutes have been readjusted. If minutes have been adjusted, seconds will stop and will not start again until the watch has been returned to the normal operating mode and the function button depressed. This allows accurate synchronization of seconds with time standards such as WWV.

There are many variations of this setting sequence, as reviewed in the Appendix on Operation. In some watches, instead of the month, date, hour, and minutes being displayed one at a time, the number to be updated is caused to flash.

In an effort to reduce costs (the corrector button does add cost to the watch case), an experiment eliminating the corrector button was done. By shifting into a set mode with five rapid pushes of the function button (within a two second time period) the various functions begin to cycle through on their own, displaying each function for 10 seconds as long as there are no further depressions of the function button. If you wish to change the hours, for example, five rapid pushes of the function button will call up the reset mode and this will cycle through month, date and then to hours. While the hours are being displayed, the function button is again depressed until the proper hour appears, then the watch will cycle on through back to normal operating mode. Whether this type of automatic setting sequency becomes popular depends upon the desires of the buyer. This is one of many experiments being tried by manufacturers to bring the best value to the customer at the lowest cost.

IC RELIABILITY

The driving force in the early development of the integrated circuit by the military was the very high reliability with which the integrated circuit could do complex jobs. During the intervening years, the failure rate of integrated circuits has been reduced to a point where only an extremely small fraction will ever show a failure, and most of them would be induced failures—failures of the integrated circuit caused by external problems.

Let's first talk about the failures that are not induced, but are present within the ICs themselves. These can be divided into two classifications. First is the defective ICs that the IC manufacturer and the watch manufacturer both failed to detect. The other is the ICs that were once good but went sour during the course of use.

First, the problems that the manufacturer fails to detect. A bit of sympathy for the manufacturer is needed. It is not uncommon to have a defect in an IC that will cause a high battery drain only at a given time on a given day. For example, at 9:00 P.M. every September 12th, it is possible for an IC to go into a high current mode of operation and discharge the battery. To detect this problem, the IC manufacturer and the watch manufacturer need to test every IC in every possible time and date and in every possible mode, at a test speed which is slow enough to pick up high current modes. Some manufacturers make an effort to do this and most failures of this type are found prior to the products being shipped to the customers. Most are found, but not all.

The ICs that fail after they leave the factory are really few and far between. ICs that fail after being tested that are not due to outside problems are a result of problems in manufacturing, and will affect most or all of the watches made from ICs of that lot. The most common problem in this uncommon classification is the change in voltage required to turn the field effect transistors in the C-MOS on and off.

As manufacturing controls become more sophisticated, the probability of this kind of problem occuring is even further reduced.

A much larger classification of IC failures is induced failures and, again, these come from two sources. One, failure to provide proper protection for the IC from moisture and physical damage, and two, static electricity. Providing protection of the IC from mechanical damage and moisture was discussed earlier. Manufacturers have all learned from prior experience and are providing protection that the integrated needs.

A much more difficult source of induced failure to control, however, is static electricity. MOS devices are by their very nature susceptible to static electricity problems. The "O" of MOS stands for an extremely thin silicon oxide layer (normally 1/5 of the wave length of light). If the breakdown voltage of this layer is exceeded, the device will be destroyed. This breakdown voltage is normally in the range of 20 to 50 volts.

Static electricity is a very potent force. When we walk across a carpet in a dry room, a spark 0.1 inch (2.5 millimeter) long between our finger and the door knob means that we have generated 10,000 volts of static electricity potential between our body and the door knob. Static electricity is generated whenever any two nonconducting materials are rubbed together. In the digital watch module, there are many such opportunities. The plastic parts used in the watch module are nonconductive. A cloth used to clean the module would also be nonconductive and could generate enough static electricity to cause the module to fail.

The problem of susceptance to static electricity has been recognized since the first MOS devices were manufactured, and all devices currently in use have static electricity protection at all input and output lines. The protection devices do not provide complete protection, and even with the best available protection, a 10,000 volt spark will generally cause a device failure. Static electricity suppression while servicing the module will be discussed extensively in Chapter 10.

BOUNDARIES OF INTEGRATED CIRCUIT CAPABILITY

There is no doubt that the horizons of the wristwatch will be greatly expanded. In the relatively brief history of electronic timekeeping, we have already reached the calculator watch. A discussion of integrated circuits for watch applications would not be complete without discussing future capabilities and limitations.

Performing digital logic (in a reasonable amount) is simple for the IC, so counting the vibrations of quartz and computing the time is a natural—as is adding a calculator function (the keyboard represents the problem, not the IC).

In a group of applications that would be more difficult but still possible are those jobs requiring audio amplification for various types of communications, or vocal announcement of time, and voltage to digital converters, as might be used for digital indication of temperature or barometric pressure. Problems for which there are not good integrated circuit solutions include radio frequency selection, or tuning, as required to make FM and AM receivers.

Within these constraints, though, are many jobs that can be effectively performed by the integrated circuit in a digital watch; the addition of a memory IC would allow the daily list of appointments to be entered into the watch and played out on request. Further use could be made by combining the watch functions and the memory so that birthdays, anniversaries, and important engagements could be stored and then remind the user of the approaching dates. The problem to be solved here is a means of inputting this information.

The digital electronic watch has brought new vitality into the watch manufacturing business and is bringing new products to the consumer for his consideration.

4

The Watch Battery

A POWER SOURCE FOR THE DIGITAL ELECTRONIC WATCH

Batteries of different size, shape and functions surround us every day, yet their operation, to an extent, is a mystery. A *battery* is a device that produces an electrical current by a chemical action. There is some evidence that crude batteries were used over 2000 years ago, but batteries used in digital watches have their technical origin beginning during World War II.

The operation of batteries used in digital electronic watches can be best explained by referring back to the Voltaic cell, shown in Figure 4-1.* The voltaic cell consists of two plates of different materials that are dipped into a chemical solution. The plates are called electrodes and the chemical solution is referred to as the electrolyte. Usually in the Voltaic cell, the negative electrode is zinc and the positive electrode is carbon. The electrolyte is a weak sulfuric acid solution. If the voltage from this cell is measured with a volt meter that draws no current, it will indicate approximately 1.8 volts. It was found that this voltage is associated with the use of these particular electrode materials and is reasonably independent of the material used as an electrolyte. Further experimentation in the time of Volta resulted in finding that different materials have different electrochemical voltages. When any two materials are immersed in the same electrolyte, the voltage developed between them is the difference in their electrochemical potentials. Table 4-1 gives the electrochemical potentials for some of the more

*Named in the honor of Alessandro Volta, 1745–1827, the Italian physicist who developed it.

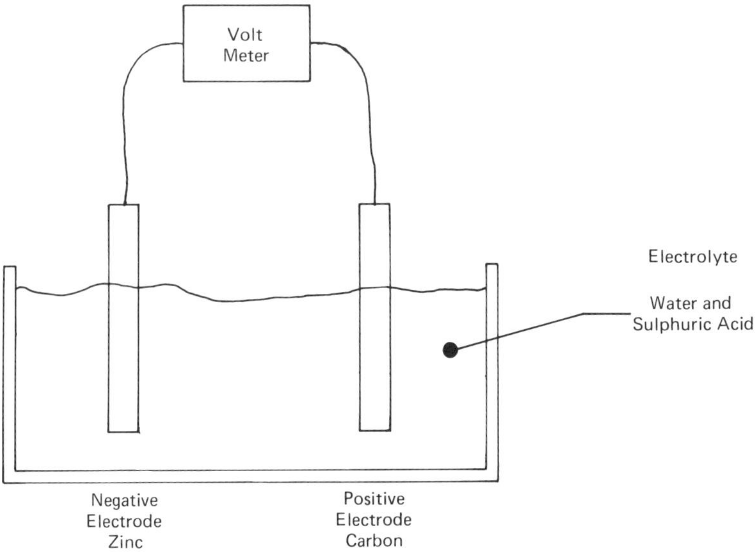

Figure 4-1. A voltaic cell.

common materials. This table is sometimes referred to as the Galvanic Series where lithium, magnesium, and aluminum are the materials which corrode most when immersed in an electrolyte in combination with other materials. On the other hand, gold and platinum are the least effected by immersion in electrolyte in contact with other materials.

Any liquid that will carry an electric current through ions can be considered an electrolyte. This group includes acid solutions, beer, salt water, and in general, any liquid that has a corrosive effect (it is interesting to note that tea and coffee are not electrolytes).

A demonstration of what can be done with a simple battery has been provided by Microma. Figure 4-2 shows a Microma liquid crystal watch being powered by a battery consisting of two limes and an orange with iron nails and copper wire electrodes imbedded in the fruit; this connected in series. Such a battery source will power a liquid crystal watch for two or three weeks (or until the fruit rots).

The technology that has gone into battery development the last 200 years has been devoted to solving the problem of anode contamination of the electrolyte, electrolyte contamination of the electrodes, and

Table 4-1. Electrochemical Voltage for Common Materials.

Element	Volts	Ions
Lithium	2.95	Li^{+}
Potassium	2.92	K^{+}
Magnesium	2.40	Mg^{+2}
Aluminum	1.70	Al^{+3}
Manganese	1.10	Mn^{+2}
Zinc	0.76	Zn^{+2}
Chromium	0.557	Cr^{+2}
Iron	0.44	Fe^{+2}
Cadmium	0.40	Cd^{+2}
Nickel	0.23	Ni^{+2}
Tin	0.13	Sn^{+2}
Lead	0.12	Pb^{+2}
Iron	0.04	Fe^{+3}
Hydrogen	0.000	H^{+}
Copper	−0.344	Cu^{+2}
Copper	−0.47	Cu^{+}
Silver	−0.79	Ag^{+}
Mercury	−0.79	Hg^{+2}
Lead	−0.80	Pb^{+4}
Palladium	−0.82	Pd^{+2}
Platinum	−0.86	Pt
Gold	−1.36	Au^{+3}

disruptive effects called polarization that stops the flow of current due to insulating layers forming on either of the anodes. Prior to the advent of the lithium battery three types of batteries have been used by electric watches; a carbon zinc battery (using a chemical system not grossly different than that described for the Voltaic cell), a mercury cell, and a silver cell. Without doubt, the silver cell has been most extensively used in digital electronic watches. Mercury cells as well as silver oxide have found use in the tuning fork and analog quartz type watches, and the older carbon zinc cells were used in the first battery operated watches that were manufactured. The batteries commonly referred to as mercury batteries are mercury-zinc battery systems, and the batteries referred to as silver batteries are silver-zinc systems. Therefore, all batteries discussed in this portion of the chapter are members of the zinc family. The broad spectrum of batteries that fall

Figure 4-2. A watch being powered by a simple battery. (*Courtesy* of Microma Corporation.)

into the zinc family should be remembered when we discuss another battery family having perhaps a wider diversity of members—the lithium family.

This description of battery construction will center on the silver cells due to their broader use in digital electronic watches. The construction of the mercury cell is very similar. As shown in Figure 4-3, a stainless steel cell can has a pellet of silver oxide as cathode material placed in the bottom. On top of this is an insulating but porous separator and an absorbent pad containing an electrolyte. The anode consists of a mixture of zinc, a corrosion inhibitor, electrolyte, and other chemical agents. Over this is placed the anode cap and insulating seals. The sides of the can are then crimped, completing the battery.

The silver oxide cathode and the zinc mixture anode are generally not contributors to battery failure. Problems with watch batteries are generally due to the barrier separator system and problems with the electrolyte. The two types of electrolytes used in these batteries are

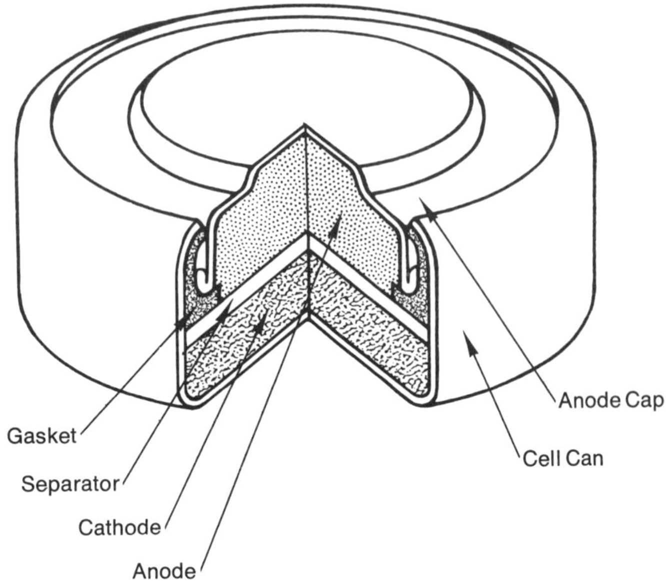

Figure 4-3. Cutaway view of a silver oxide battery. (*Courtesy* of Union Carbide Corporation.)

pasty solutions of either sodium hydroxide or potassium hydroxide. Sodium hydroxide is sold in dry form as the plumbing cleaner Drano, and potassium hydroxide is even more wicked.

While these materials make excellent electrolytes, they share a problem. They tend to creep along metal surfaces; thus, the sealing of these materials into a watch battery has been a problem that has constantly plagued battery manufacturers. The effect of this creeping can be observed on the top of the cells as either a free liquid, or dried salt crystals, which result from the reaction of the electrolyte with carbon dioxide in the atmosphere. The presence of either of these forms of creepage is considered not acceptable in an electronic watch, and should be cause for battery replacement.

The battery, when new, has a theoretical voltage of 1.6 volts, but when power is drawn from the battery, the voltage goes to approximately 1.5 volts and stays there throughout the life of the battery. A curve typical of this kind of battery performance was shown earlier as Figure 2-15.

COMPARISON WITH HEARING AID BATTERIES

Quite a few of the watch batteries are identical in overall size and configuration with batteries manufactured for hearing aids. The use of hearing aid batteries in electronic watches is not recommended. These batteries are designed to power hearing aids for one or two weeks at current drain significantly higher than those found in electronic watches. The need for higher current output and the tolerance of shorter life combine to make the hearing aid battery different than its digital watch counterpart.

The potassium hydroxide electrolyte used in hearing aid batteries contributes to its higher discharge rate capability, but also causes difficulty in sealing problems, especially over long periods of time. The barrier film used in hearing aid batteries provides adequate protection from internal shorting during the hearing aid battery life, but multiple barrier layers are required for the longer life of the digital electronic watch.

To summarize, if a hearing aid battery is used in an electronic watch, more than likely the life of the battery would be significantly less than that available from the proper watch battery, and more important, possible leakage from the battery during this extended duty might cause serious damage to the module.

Just for note, batteries that are designed for use in LED watches can be used in hearing aids without any dire consequences. They will not have a longer life than a hearing aid battery, but they should provide adequate service.

HIGH DRAIN BATTERIES FOR LED WATCHES

As will be described in Chapter 6, the drain on a battery from a LED module can be as high as 80 milliamperes during display of time. This compares with 5–8 microamperes (1/1000 of a milliampere) required by the liquid crystal watch during normal operation. This diversity of requirements, even on cells of the same voltage and physical size, has resulted in high drain-rate cells being produced for the LED, and low drain-rate cells being produced for the LCD. Table 4-2 gives a comparison of some of the more common cells in low drain-rate and high drain-rate classifications.

Table 4-2. Low and High Drain Rate Batteries.

LOW DRAIN SILVER BATTERIES:				
Size		Battery Number		
(in.)	(mm)	Eveready	Mallory	Ray-O-Vac
.455 × .160	11.56 × 4.06	301	WS11	RW14 or 34
.455 × .220	11.56 × 5.59	303	WS14	RW12 or 32
.310 × .210	7.87 × 5.33	309	10L13	RW18 or 38
.310 × .143	7.87 × 3.63	384	10L15 or WS10	RW37

HIGH DRAIN SILVER BATTERIES:				
Size		Battery Numbers		
(in.)	(mm)	Eveready	Mallory	Ray-O-Vac
.605 × .190	15.37 × 4.83	355	10L129	RW25
.455 × .210	11.56 × 5.33	357	10L14	RW22
.455 × .165	11.56 × 4.19	386	10L124	RW24 or 44
.310 × .140	7.87 × 3.56	392	10L125	RW47
.310 × .210	7.87 × 5.33	393	10L123	RW48

The principle difference between these two classifications of batteries is the electrolyte. The low drain-rate contains a sodium hydroxide electrolyte. This is a relatively docile material and does not have the electrolyte leakage or salting problems found with potassium hydroxide, the electrolyte used in high drain batteries.

Some manufacturers believe that the seal problem with potassium hydroxide is solved, and the high drain-rate batteries are usable with both high drain and low drain watches. This is truly one of the subjects where time will tell. A "family portrait" of watch batteries is shown in Figure 4-4.

MORE ENERGY STORAGE IN SILVER BATTERIES

In the past, all silver oxide batteries were made of silver oxide in the form of Ag_2O. Recently, a silver oxide battery was manufactured which uses divalent silver oxide (AgO) as the cathode material. This theoretically results in more energy storage in the same battery

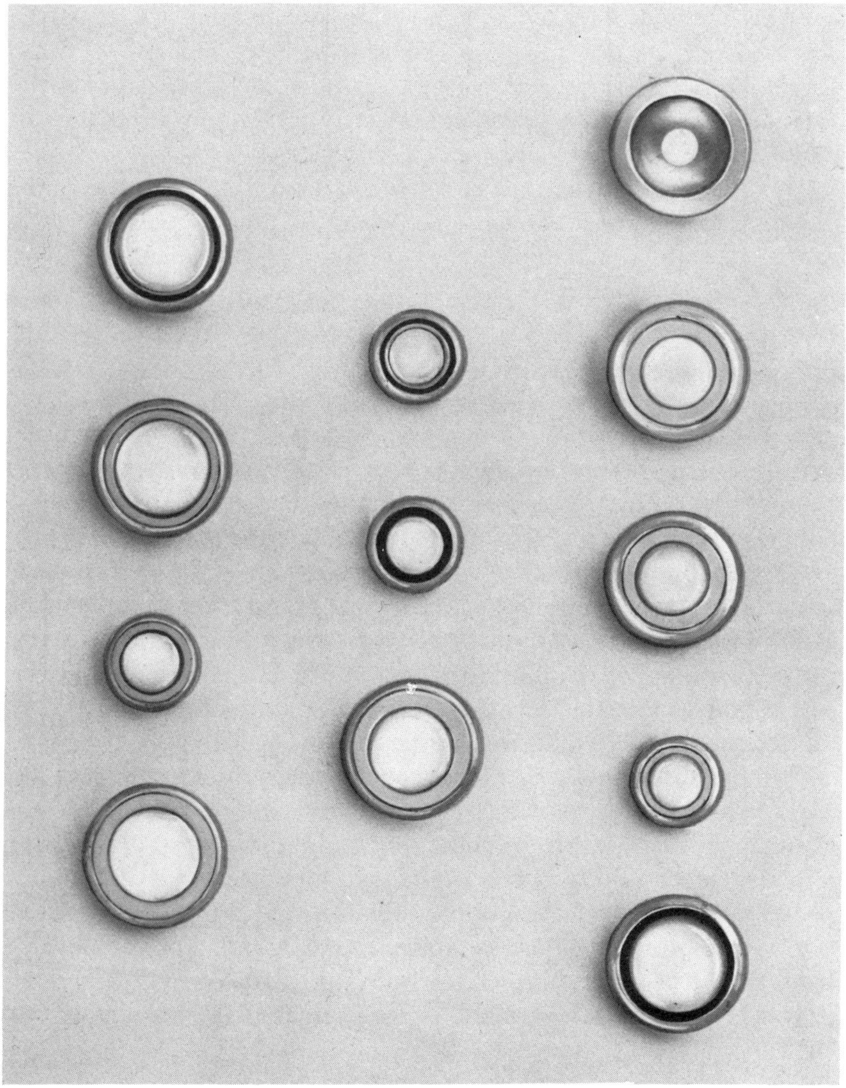

Figure 4-4. Various batteries used in watches. (*Courtesy* of Union Carbide Corporation.)

volume. Manufacturing control problems have prevented much of the divalent silver oxide batteries potential from being realized. Tests on these batteries during the early years of production showed that on the average, the power they stored was not significantly more than that for the more normal Ag_2O silver oxide batteries.

There is no question that the divalent silver oxide does have the potential for 10 to 15% greater energy storage, and as manufacturing processes mature, the expectation of a somewhat longer life silver oxide battery should be realized.

LITHIUM BATTERIES FOR DIGITAL ELECTRONIC WATCHES

Batteries using lithium as an anode material form a system that is in competition with those that use zinc as an anode material. For these batteries to provide viable competition for the more thoroughly developed zinc based batteries, there must be distinct performance advantages for the lithium batteries. The performance advantage available depends, of course, on the type of problem that is to be solved. For the LED watch, the availability of 3 volts directly from one battery solves a distinct problem. For the liquid crystal watch, 3 volts might mean that a voltage converter is no longer required. Another advantage might be a longer life through improved energy storage and reduced self-discharge, which gives longer shelf life.

At the time of this publication, lithium batteries are still an immature technology. Various manufacturers have developed approximately a dozen lithium cell systems for watch use. These lithium cells are as different from one another as the carbon-zinc voltaic cell is from the silver oxide-zinc cell for watch applications. The characteristics which are most important for the electronic watch are as follows.

1. Energy density—some of the lithium systems have somewhat higher energy density than is available with the silver oxide-zinc systems; others have lower energy density but do have other advantages.
2. Self-discharge—most batteries in use today have a continuous discharge due to an internal chemical imbalance. Since this discharge is present even when the battery is not in use, it limits the shelf life of the battery (most silver oxide batteries can have

shelf lives of two to four years). The self-discharge of most lithium systems is believed to be well below that of the silver batteries used in watches today, and therefore can be depended upon for longer shelf life and longer operating life on very low drain watch modules. Not all lithium systems share this characteristic.

3. Terminal Voltage—terminal voltage is important because approximately 3 volts is required for operating an LED watch. However, minimum power for operating the watch circuitry occurs at as low a voltage as the circuitry can reliably operate. Most IC manufacturers peg their desired operating minimum voltage at 1.5 volts. Operating at voltages higher than this value results in excess power consumption.
4. Discharge Curve—the discharge curve of a cell indicates how the terminal voltage of a cell changes during the life of the battery. silver cells, for example, have an extremely flat voltage curve over the life of the cell. This lack of change is important because the timekeeping accuracy changes as the battery voltage changes. Many lithium batteries, on the other hand, have either a stepped discharge curve, or a continual slope during the life of the battery. With most integrated circuits, this significantly degrades the accuracy of the watch.

Figure 4-5. Lithium battery featuring long shelf life.

Lithium batteries made their debut in digital watches manufactured by several organizations during 1977. Each manufacturer had a different purpose in mind for the introduction of the battery. The battery shown in Figure 4-5, introduced by Micro Display Systems, features extremely low self-discharge and therefore very long shelf life. This, when combined with very low-drain electronics, yields an expected battery life longer than 10 years. Other lithium battery systems introduced by LED watch manufacturers extend the battery life of the LED to two to three years and require only one battery. It will be many years before the relative position of the lithium and the zinc based battery systems in electronic watches have been determined.

NOTES ON BATTERY HANDLING AND REPLACEMENT

Battery Life

Only the liquid crystal watch without backlight (or with tritium backlight) represents a constant load to the battery that is not subject to the whims of the customer. Even in these watches, manufacturing tolerances run as much as two to one current drain even between watches in the same lot. Therefore, for one watch to obtain a 24 months lifetime and one identical to it to obtain a 12 months lifetime, is completely normal. When there is a backlight in an LCD, or in an LED where customer utilization determines the battery life, the problem is even more complex. The experience of LED manufacturers has been that a customer (in the early days of the LED, especially) would wear out the first set of batteries in three months and the second set in six to nine months, and after that, the batteries would easily meet their rated 12 months life expectancy. This was obviously not the battery manufacturer's fault, but involved customer use patterns.

Solar Cells for Digital Watches

Some digital watches have been built with semiconductor solar cells to recharge the batteries. These silicon devices are located on the face of the watch, under the crystal, as shown in Figure 4-6. Each individual device generates 0.3 to 0.5 volts at a few microamperes to a few

Figure 4-6. LED watch having ten solar cell panels on its face.

milliamperes, depending upon light level.* In very low light levels, they generate no power and batteries are required for operation.

The use of solar cells in digital watches can be questioned on technical grounds, but customers seem to accept their usefulness. The problem is that rechargable batteries, even if they are recharged, do not have a longer life than the shelf life of other battery types. On LCD watches, the power consumption is low enough that the shelf life of the batteries is approached. On LED watches, solar cells do contribute to extending battery life.

Recharging Watch Batteries

With solar cell recharging being used occasionally on both LED and LCD watches, some watch batteries have been designed so that

*Four or five are connected in series and generate the 1.5 volts needed to supply power to the watch and recharge the batteries.

recharging is possible. Special silver oxide batteries have been developed for recharging, and also nickel-cadmium batteries have been built in watch battery sizes. These are the only types that should be used in the solar cell watches. Watch batteries not made for recharging should not be recharged. This is because of two reasons: 1) internal pressures can be developed which can cause the battery to rupture, and 2) due to the chemical systems involved, only a minor amount of actual recharging is possible.

New Battery Storage

After considerable confusion, even on the part of watch battery manufacturers, as to the best way to prolong the shelf life of the battery, an answer is now available. Watch batteries store best at room temperature (about 70° F) in a dry place. The efforts on the part of battery manufacturers to prolong the shelf life by either freezing or storing them just above freezing temperature have not proved worthwhile. Consequently, chilled storage facilities installed at watch battery manufacturers are no longer being used.

It is also desirable to keep your supply of watch batteries in their original packing to avoid accidental shorting. They should not be stored in trays with other batteries or other metal parts, due to the possibility of short circuiting.

Battery Interchangability

Batteries produced by the various manufacturers can be interchanged as listed in each manufacturer's interchangability guide. The fact that a battery is the same physical size does not imply interchangability. It could be a hearing aid battery of the mercury-type (which has a lower voltage and therefore will not function properly). The battery replacement guides produced by all watch battery manufacturers should be your guide in battery replacement.

Battery Replacement

If the batteries removed from a watch show any signs of salting or electrolyte leakage, the watch should be examined for damage and the contacts cleaned if necessary. When replacing a watch battery, be sure to insert the battery properly in the watch according to the watch manufacturer's instructions. Use only the battery (or proper replacement) listed in the replacement guide. When handling batteries, use plastic or plastic tipped tweezers. Metal tweezers will short out the battery.

Disposal of Used Batteries

Mercury batteries represent a source of ecological concern due to their mercury content; therefore, disposal of these should be given some thought. Also, silver batteries, while not representing the ecological problem that mercury batteries do, do represent a realizable cash value. The small quantity of used batteries collected by the individual jeweler and watchmaker do not generally justify a reclammation program by the jeweler or watchmaker, but the American Watchmakers Institute at 3810 Harrison Avenue in Cincinnati, Ohio 45211 is collecting these batteries, and the proceeds will help fund its scholarship program.

Figure 4-7. Used watch batteries do have cash value. (*Courtesy* of Union Carbide Corporation.)

5
The LED Display

In the period from 1955 to 1965, there was an absolute explosion in knowledge of semiconductor materials and devices. During this 10-year time period came the present versions of transistors, integrated circuits and other semiconductor devices. Most transistors and integrated circuits are now made with silicon, but a large amount of research on other materials—especially gallium arsenide—was done. In 1961, researchers at Texas Instruments found that a gallium arsenide diode or transistor when conducting current gave off an infrared light. As this phenomenon was better understood, scientists learned how to modify the gallium arsenide material so that it would produce light in the visible region instead of infrared.* A visible light emitting diode was first demonstrated by researchers at General Electric Company in 1962.

Light Emitting Diode (LED) displays were available as early as 1964, but were handmade and very expensive, costing up to $10 per digit. The first digital watch was manufactured for Hamilton Watch Company by Electro-Data of Dallas, Texas and used LED displays.

A gallium arsenide device generates only a spot of light and many spots are needed to generate a character. The first LED digital displays consisted of 20 to 30 devices in the traditional figure 8 pattern as discussed in Chapter 1. These were connected so they could be turned on in groups to form characters. The next improvement was to make several diodes on the same piece of gallium arsenide so that these longer pieces could be located, again, to form a figure 8. Only one

*The modified material is actually gallium arsenide-phosphide, but we will refer to it as gallium arsenide in this book.

Figure 5-1. Hybrid display digit.

connection was required to each piece. This is referred to as a hybrid display. A character of this type is shown in Figure 5-1.

When current is applied, the dark portion of the bar lights up, and holes in the lighter colored metal contact pattern on top of the bars allow light to shine through the metal pattern as well as from the edges of the bar. Figure 5-2 shows a complete display that is encapsulated and ready for use in a digital electronic wrist watch.

Hybrid construction has several problems—which total to be a cost problem. It is difficult and costly to handle and align each of the seven bars for each digit. The eye can perceive very minute misalignments in each character. There is also considerable possible variation in the efficiency (and therefore light output) of the various LED bars. The bars that form a character and a display must all be closely matched in their light output characteristics. Again, this is a cost factor. The third

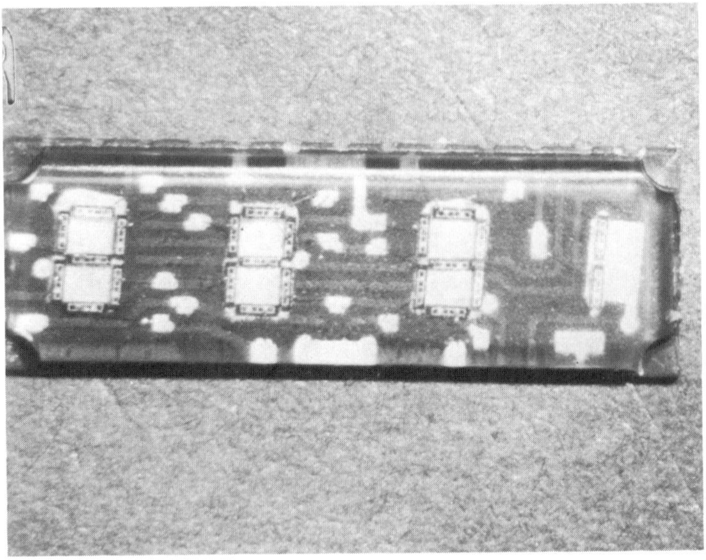

Figure 5-2. Hybrid display ready for use in LED watch.

and most obvious cost factor is the difficulty in handling the large number of pieces of such small size required to build a watch display. There needed to be a better approach, and the monolithic character was it.

A monolithic display has all seven bars made as a part of one semiconductor device. This approach was not used earlier due to the high price and great fragility of gallium arsenide. During the early 60's, the price came down and handling techniques were improved, so that monolithic digits up to 0.120-inch high are feasible. Most LED watches using monolithic digits have used the 0.100-inch high size. A configuration that forms not only numbers but letters, is shown in Figure 5-3. Four such digits can be combined and encapsulated to again form a watch display, as shown in Figure 5-4.

The digits of a monolithic display are about the size of a relatively large integrated circuit, and it is not surprising that the costs are not grossly different than the cost of an integrated circuit chip. Prices of $1.00 to $1.50 per digit are not uncommon for the 0.100-inch high monolithic display. While this cost is acceptable for higher priced LED watches, it is unacceptable for higher priced LED watches, it is unacceptable for the less expensive products. The most straight-

Figure 5-3. Monolithic display digit.

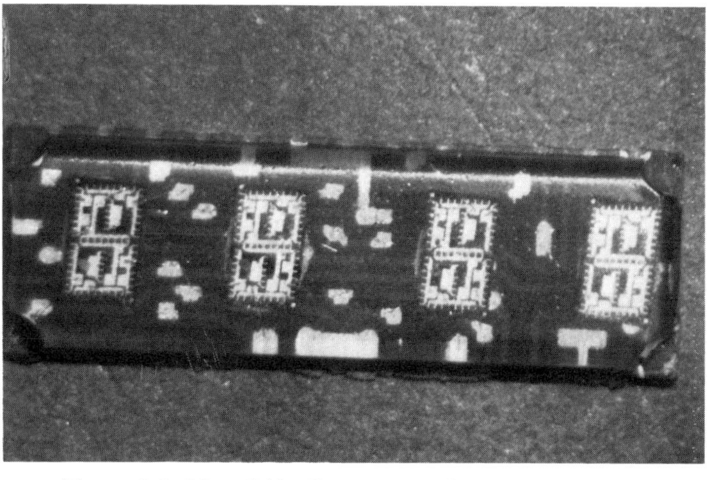

Figure 5-4. Monolithic display ready for use in LED watch.

forward way to reduce the cost of the LED display is to make the digits smaller. There has been some experimentation with digits 0.080-inch high, but this has met with some resistance on the part of the customer due to readability problems. The alternate approach, and surely the one most widely pursued, is the addition of a magnifier over each digit or in some cases over the entire display. Twice as many monolithic LED displays can be manufactured on one slice of gallium arsenide (and therefore result in about one-half the manufacturing cost per digit) if a magnifier of 1.4 power is used. Figure 5-5 shows a 0.060-inch high monolithic display with a 1.67 magnifier molded on to provide the effect of a digit 0.100-inch high. Further reductions can be obtained by going to higher powers, but there is a penalty that is paid, and that penalty is viewing angle.

The restriction in viewing angle with magnification for LED displays is both an engineering and a judgment question. Some

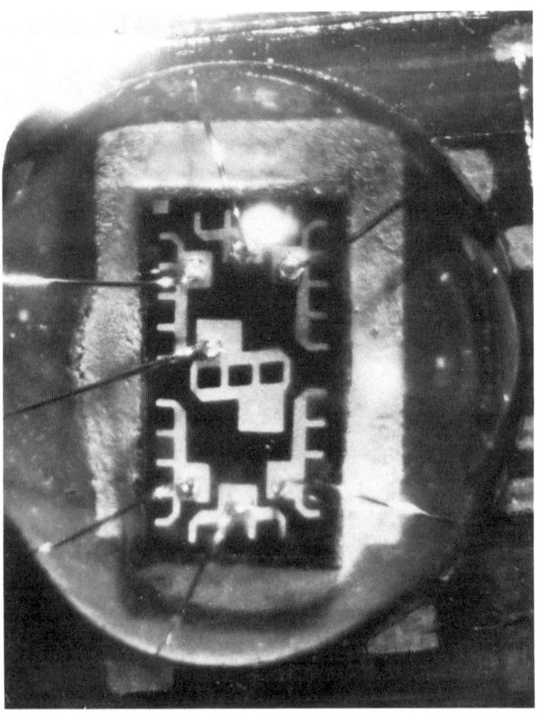

Figure 5-5. Monolithic digit with molded on magnifier.

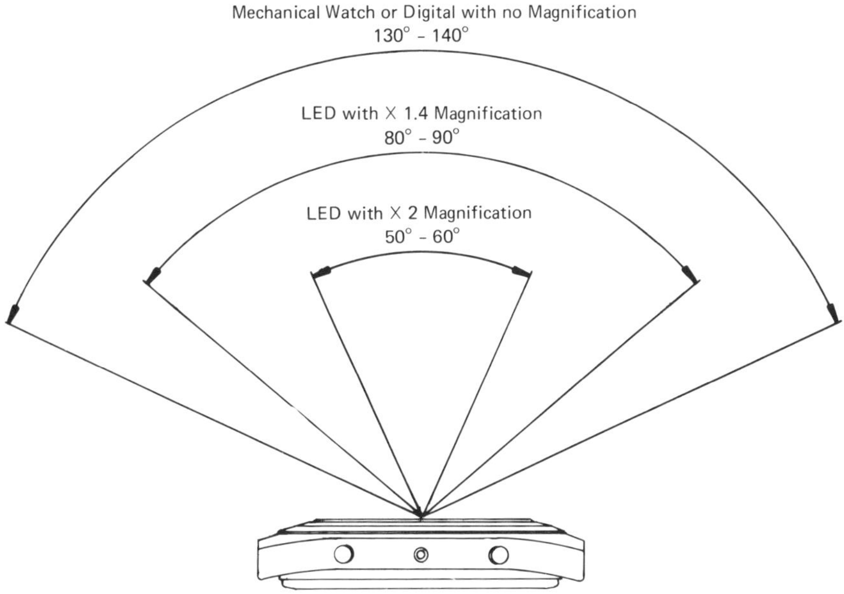

Figure 5-6. Effect of magnification on viewing angle.

magnification systems give better performance than other magnification systems. For example, molding the magnifier directly on to the display substrate provides a broader viewing angle than does using an external magnifier. Also, what constitutes viewing angles is strictly viewer judgment. Even so, it is worth while to give an approximate reduction of viewing angle with increased magnification. The face of a mechanical watch, for comparison, can normally be read over a viewing angle of 130 to 140°, as shown in Figure 5-6. An unmagnified LED or a well designed liquid crystal display, also has a similar viewing angle. If, as in the earlier example, we use a 1.4 magnifier over a smaller LED character, then the viewing angle is reduced to the order of 80 to 90°. If another reduction by a factor of two in the LED bar size is desired (and this is being used on the lowest cost LED watches), then the viewing angle is further restricted. A 0.050-inch high digit with a times two magnifier will give a viewing angle to the order of 50 to 60°.

The power required by the LED display impacts the size of the batteries, and the size of the watch, and determines how many times the display can be turned on before the batteries must be replaced. The

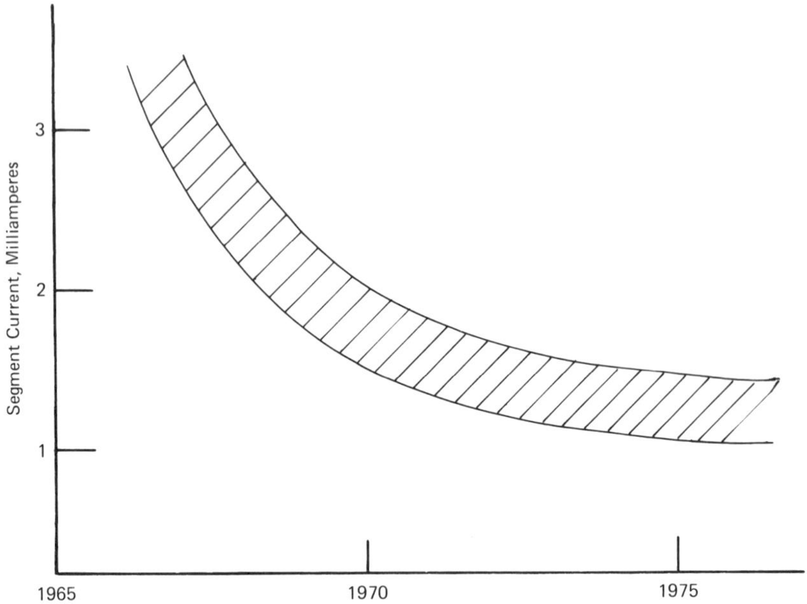

Figure 5-7. Trend of LED efficiency.

power required by LEDs has been reduced dramatically since their invention. Figure 5-7 shows a general trend line, with very rapid reductions in power consumption for a given brightness (during the late 60's and early 70's). However, it is obvious that there has been very little progress in the last few years and no great amount of progress is expected since most of the earlier efficiency losses of the LED display have already been eliminated. No major reduction in current consumption of the LED is forecast by major manufacturers at the present time.

A newer semiconductor material, gallium phosphide, had been expected to solve most of the problems of the LED. Gallium phosphide had the ability to make not just one color displays like the deep red of the gallium arsenide, but to make many colors, from red through green, as determined by the exact material composition. Great things were expected for this material because its theoretical efficiency is higher than that of the gallium arsenide. To an extent, higher efficiency has been achieved, but it has been found that the effective visibility in sunlight is, at best, no better and in most instances worse than the visibility experienced with gallium arsenide. This is because gallium phosphide puts out light over a comparatively broad spectrum, whereas gallium arsenide puts out light of almost a single pure color. The very deep red of the gallium arsenide is contrasted with the much softer reds, yellows and greens of gallium phosphide. The softer colors prevent the use of filters as efficient in reducing the sunlight problems as used with the gallium arsenide displays.

As was the situation with the gallium arsenide, only minor advances are expected to be made in gallium phosphide technology during the coming years. Gallium phosphide will be used in a few digital watches, but will not replace the workhorse status of gallium arsenide.

NEW TRENDS IN LED DISPLAYS

LEDs are rapidly approaching the status of a mature technology; that is, major new developments are coming more slowly now. This is not to say that there is nothing new in the field of LEDs, there is. In the following paragraphs, the ways of LED capability expansion will be explained.

Color

As mentioned earlier, gallium phosphide is available not only in a red, but in yellow and green. The availability of these, plus the deep red of gallium arsenide, opens many new product opportunities. The light output of the gallium arsenide LEDs or gallium phosphide LEDs are reasonably equal to each other. However, it is easier to provide correct filtering for the deep red of the gallium arsenide LED and, therefore, the red display can be seen in brighter sunlight than any other color.

For evening wear and fashion purposes, the colors available from the LED should provide the stylist and designer with an endless variety of new product possibilities.

Alphabet

The LED characters shown in Figure 5-3 have extra connections (that do not significantly add to cost) that permit not only numbers, but

Figure 5-8. LED watch showing day-of-week.

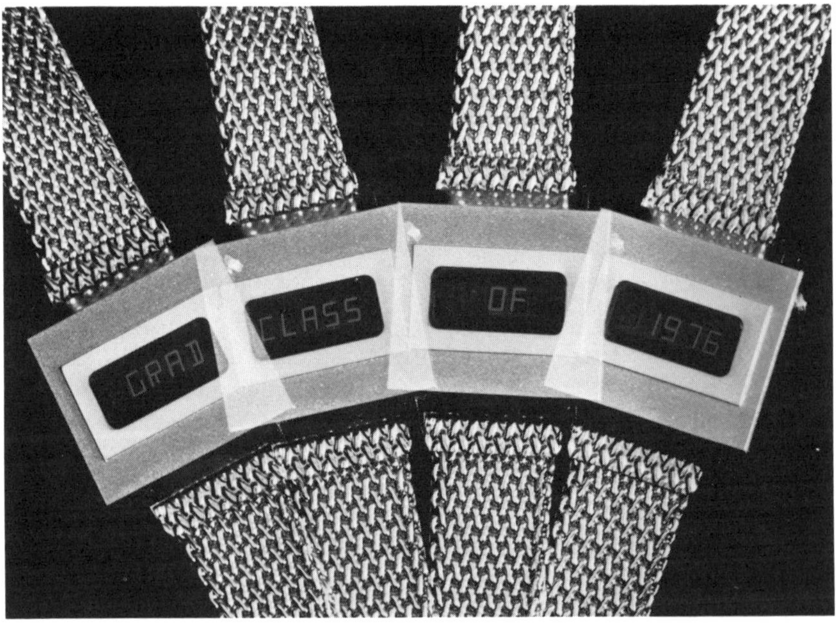

Figure 5-9. Hughes Aircraft Company message watch. (*Courtesy* of Hughes Aircraft Company.)

letters of the alphabet to be formed. This feature allowed the LED to be the first watch to show day of the week in a format shown in Figure 5-8.

The use of the alphabet capability of the LED was most completely used in a watch module manufactured by Hughes Aircraft Company. This watch module, besides showing time, can also show a stored message of five words of five letters each, as shown in Figure 5-9.

Special Characters

LEDs have been mainly used by illuminating various segments to form numbers and letters. Often overlooked is their ability to be customized for forming special logos or other characters. An example of this use of the LED technology are the day-of-the-week characters made by Hewlett-Packard for Pulsar. Figure 5-10 shows the LED character and Figure 5-11 shows the Pulsar watch that uses these display.

Figure 5-10. Hewlett-Packard LED day-of-week characters.

Figure 5-11. Pulsar LED watch using day-of-week characters. (*Courtesy* of Time Computer, Inc.)

Further into the Future

Despite the many capabilities of the LED display, the main feature of importance in the future is the low cost, especially with a magnifier, and its compatibility with semiconductor manufacturing processes. The low cost of the LED displays is self-explanatory. The compatibility of module manufacturing processes with the IC will become more obvious in the next chapter on the LED watch. The result is low cost—the dominant factor in use of the LED display.

(Note: The author wishes to thank Bob Cole and Carroll Smith of Texas Instruments for display samples and helpful discussions.)

6
The LED Watch

THE JOB OF THE LED MODULE

The LED displays discussed in the preceeding chapter are only a small part of the electronics of an LED watch. The other components required are the quartz crystal and variable capacitor, the integrated circuit to divide the quartz frequency, compute time, and turn on the necessary display segments. All these parts must be mounted on a circuit board, or its ceramic equivalent, and make contact with batteries, push buttons, and the case.

The way the LED display is driven (or turned on) requires some comment. LED displays can be turned on by simply connecting them across a flashlight battery. Forming digits this way requires one wire per segment (or seven wires per digit) plus one common wire. For the four-digit watch (actually a "1" and three full digits, plus colon), there would be 24 connections required to the display. The nature of LED displays are such that they can be "scanned" (the engineering term is multiplexed). This reduces the number of connections to both the integrated circuit and the display—from 24 to 11 for the four-digit watch and from 38 to 13 for six-digit watches.

The connection of the display in a scanned or multiplexed connection is shown in Figure 6-1. This is a connection to each digit that allows the integrated circuit to choose which digit will be turned on. Similarly positioned segments on each digit are then connected together so that, for example, to turn on the top segment of the second digit, battery voltage would be applied between terminal number 2 and terminal number A.

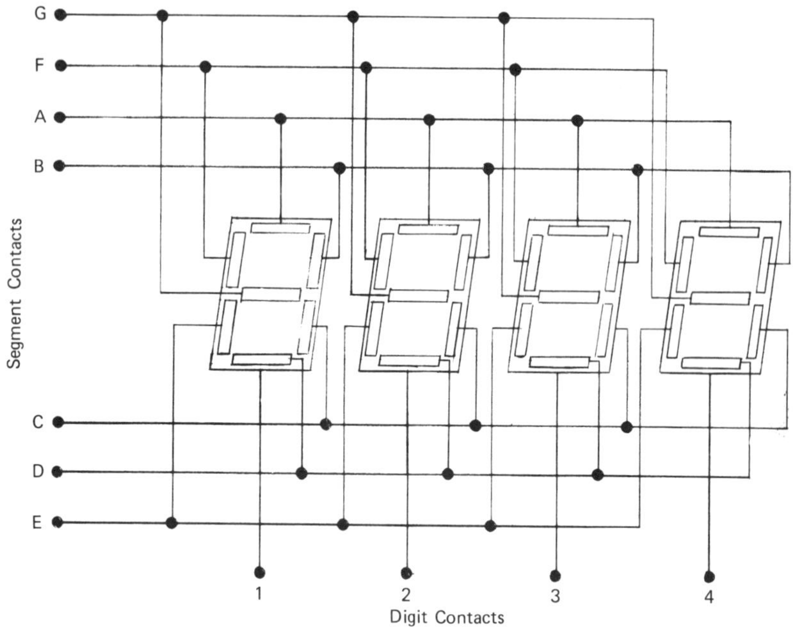

Figure 6-1. LED display connection for multiplexing.

In LED watches, the display is turned on one digit at a time. For example, the first digit (Contact 1) would be energized and voltage would be applied to contacts A through G as needed to form the desired character. This would then be repeated with digits two, three and four. Although only one digit is ever turned on at any one time, the speed of going from digit to digit is such that all digits appear to be on at the same time.

Scanning is only important to the watchmaker or user in that the reduced number of contacts on both the integrated circuit and the display serve to lower cost. The only way that scanning is apparent is if the watch is moved quickly while the display is energized. Scanning becomes apparent as the digits show one at a time like they were appearing in space. The effect is much more pronounced in a darkened room. Neither the strobing effect nor the rate at which the strobing occurs has anything to do with the quality or performance of an LED watch.

The electrical interconnection (block diagram) of an LED watch is shown in Figure 6-2. The quartz oscillator is powered by, and controls

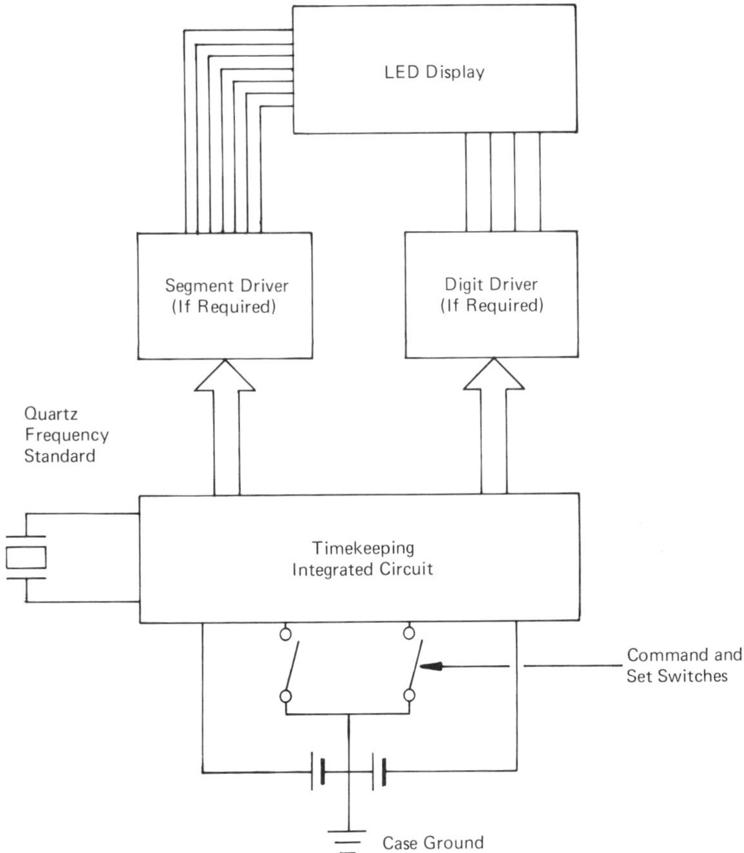

Figure 6-2. Block diagram of LED watch.

by the integrated circuit, which divides the quartz frequency to one pulse per second. Further divides calculate seconds, minutes, hours, and date. This information is then decoded. That is, the LED segments to be turned on are controlled by the count of hours and minutes when the display button is pressed to show hours and minutes.

The block diagram shows segment drivers for the seven segment lines and digit drivers for the four (or six) digits. These drivers can consist of either 11 individual transistors or two or three integrated circuits separate from the timekeeping IC or, in some recent watches, these drivers can be a part of the timekeeping IC. Connection to the two batteries and the set switches complete the electronic circuits.

BATTERY LIFE IN AN LED WATCH

The LED is a light generating display—it makes light—whereas the LCD (not including backlighting) only reflects available light. The very fact that the LED display does generate light means that it consumes power, much more power than displays that only operate from available light. Therefore, one of the most important considerations in the LED watch is the sources and uses of power. All digital watches are powered by small batteries, and the amount of power that these batteries contain is a function of the physical size of the battery. A large battery can obviously store more power. Table 6-1 gives the size and approximate power storage of a few of the most popular batteries used in LED watches. The power stored in the batteries varies from over 200 milliampere-hours for batteries that might be used in a large men's watch, down to under 40 milliampere-hours for batteries that might be used in a small ladies watch. Battery problems are compounded by the fact that both gallium arsenide and gallium phosphide LEDs require more than 1.5 volts for their operation. Therefore, two batteries are required.

Voltage converters are used in some LCD watches to provide the higher voltage required. However, the size of such a converter for LED watches would be larger than the size of the watch due to the high current required, and is therefore, not practical. As long as the present silver battery system is used, two batteries will be required. Some of the lithium batteries discussed in Chapter 4 produce 3 volts. Only one of these batteries would be required to power an LED watch.

Assuming the use of one of the more common batteries used in LED watches, with dimensions of 0.455 inch in diameter by 0.165 inch high, approximately 120 milliampere-hours energy storage is available. In the LED watch, two batteries are connected in series such that their voltages add. The resulting energy storage with two batteries present is 120 milliampere-hours at 3 volts.

Table 6-2 gives the power consumption calculations for an average LED watch to achieve one year operation from a pair of batteries. The

Table 6-1. Common LED Watch Batteries.

Size	Milliampere-Hour Capacity	Union Carbide Number	Ray-O-Vac Number	Mallory Number
.605 × .190	250	355	RW25	10L129
.455 × .210	180	357	RW22	10L14
.455 × .165	120	386	RW24 RW44	10L124
.310 × .140	38	392	RW47	10L125
.310 × .210	75	393	RW48	10L123

Table 6-2. LED Power Consumption Calculations (for one year operation).

Power Source: 2–100 milliampere-hour (mA h) batteries connected in series to provide 100 mA h at 3 V.

IC Power Consumption: 5 μA. In 1 yr (8760 hr), the IC will consume 43.8 mA h.

LED Power Consumption: Each segment draws one mA. There are 26 segments in a watch display, but only an average of 14.6 are on at any one time.

One Year Operation:		
	Battery capacity	100 mA h
	IC for 1 yr	43.8 mA h
	Remaining	56.2 mA h
	LEDs consume	14.6 mA h/hr
	LED use before battery discharge	$\frac{56.2}{14.6} = 3.85$ hr/yr
	LED average use per day	$3.85 \times \frac{3600}{365} = 38$ sec/day

power source is assumed to be two 100 milliampere-hour batteries connected in series to provide 100 milliampere-hours at 3 volts. The IC operating at 3 volts consumes an average of 5 microamperes current drain. Therefore, in one year, the IC themselves will consume 43.8 milliampere-hours, or just under one-half of the energy available in the battery. The big culprit, though, is the power drawn by the LEDs themselves. Each segment draws one milliampere average current to provide adequate viewing brightness. An average of 14.6 segments are required to show time; therefore, the viewing time for one year's operation can be readily calculated. The batteries provide 100 milliampere-hours power; the IC require 43.8, and 56.2 milliampere-hours is left over for providing power to the LEDs. If the LEDs require 14.6 milliampere-hours per hour of display operation, then the LEDs can only be operated for 3.85 hours per year. Reducing this from hours per year to seconds per day gives 38 seconds per day, or something over 30 one-second viewings of the time per day.

Setting an LED watch causes battery drain only to the extent that the time the display is on subtracts from the 3.85 hours that batteries can power the display. There is no significant power drain of the IC that results from the setting operation.

Thirty viewings of the time per day is adequate for some people, and completely inadequate for others—depending of course on job and habit. Is improvement possible so that there can be continuous display? Let us look at the power consumption of each of the components. The IC at 5 microamperes can be reduced and reduction by a factor of two is definitely in the cards during the next five years. The possibility of LED improvement was discussed in the preceding chapter. No major improvement is expected in the foreseeable future. To even get a factor of two improvement from one mA per segment used in the calculations would require both some semiconductor improvements and increasing use of magnification or other light collection devices.

Batteries are subject to considerable discussion. Silver batteries are being improved, and batteries with increased capacity (10 to 20% more) are being produced at the present time. Significant improvements beyond this are not expected. An alternate type of battery which produces 3 volts directly, the lithium battery, is being introduced. Going to lithium will reduce the number of batteries required to one, but the one lithium will have as large a volume as the two batteries previously used. Due to improved use of space, there is a possibility for a times two improvement in the battery energy available. A lithium battery watch would have 200 milliampere-hours battery capacity.

If all of these improvements come to pass over the coming five years, it would increase by a factor of five or so the length of times the LED display can be shown. Much more likely, the physical size of the batteries would be reduced to make a smaller watch.

The LED watch battery situation can be placed in better perspective by considering the size battery required for one year operation with the LED display on all the time, with an average of 14.6 milliampere-hours per hour required by the LED, and 8760 hours per year. This would yield a total of 128 ampere hours required at 3 volts. The average car battery stores 75 ampere hours at 12 volts; therefore, a battery approximately half the size of a car battery would be required for one year operation of an LED watch.

THE PROBLEM IN THE SUN

Tied in with the battery power consumption problem is an associated problem: readability of the LED watch in sunlight. The basic problem is that the sun is very bright. The standard of brightness used throughout the industry is the foot candle. This unit was developed as the brightness one foot away from a standard candle. The illumination encountered in homes is of the order of 50 foot candles, and in the normal office, approximately 70 foot candles. Concentrated lighting for very fine detail work might bring this up to 300 or 400 foot candles. If we go outside, it's an entirely different world. Normal direct sunlight is in the order of a few thousand foot candles. Measurement on bright days in areas of snow or bright sand can yield reading of over 10,000 foot candles.

The usual LED watch works well in lighting of 75 to 100 foot candles, which is normal indoor light. Going outside on a dull day, a minor amount of shading will allow easy visibility with lighting up to several hundred foot candles. But, bright days are a factor of 10 brighter than this, and the optical power is just not available for the type of electronics and batteries used in LED watches. None of the components commonly used in LED watches, neither the battery, the IC, nor the display, are built now or probably can be built in the future to provide the current and the light intensity that would be required for complete outside visibility. Some LED watches are built with brightness sensors that automatically increase the brightness of the LED in a bright environment, but this increase is only a factor of four, or sometimes a factor of eight, and is still a factor of 10 less than the brightness required to be easily visible outside all the time.

While the power cannot be increased sufficiently to give outside visibility under all conditions, measures can be taken help solve this problem. If the filters have been properly designed by the manufacturer to match the characteristics of the LED display, further filtering cannot increase the outside visibility.

It is obvious that the display should be shaded from the sun. Shades can be manufactured as a part of the watch, as shown in Figure 6-3. (The large areas on the top of the watch are for solar cells to recharge the batteries.) For the more common configurations of the LED watch, such shading is inconvenient and bulky. A comparable shading can be obtained through the use of magnifiers. If the sun is outside of the field of view of the LED display, then the sun cannot illuminate the

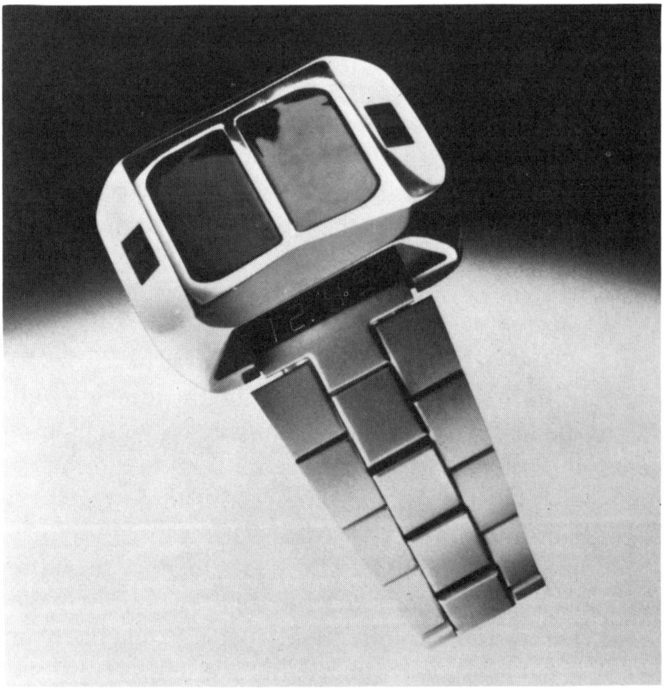

Figure 6-3. LED watch with shaded display. (*Courtesy* of Ragen Information Systems.)

characters of the LED watch through the red filter and outside visibility is significantly improved. An LED watch using magnifiers has significantly improved outside viewing compared to one using hybrid or unmagnified monolithic displays, even though the light output of the two might be the same.

LED MODULE CONSTRUCTION

There are many paths to the use of an LED display in a watch. One popular LED module is that manufactured by Hughes Aircraft Company, shown in Figure 6-4. In this module, the four LED displays and the integrated circuits are mounted and bonded directly to a ceramic substrate. The four black objects in line across the center are the LED displays themselves. The large rectangle near the bottom is the C-MOS integrated circuit that calculates times.

The displays on this Hughes module are scanned (or multiplexed), as discussed in a preceding section. Careful examination of the photograph will show that all of the similar segments of each display are connected to the same gold line on the ceramic substrate. Near the C-MOS integrated circuit is a smaller integrated circuit that serves as a segment driver. It takes low current outputs from the C-MOS IC and drives the segments at a much higher current level than the output that this particular timekeeping IC can provide.

Similarly, the integrated circuit off by itself to the right of the C-MOS is the digit driver. It takes four digit outputs of the C-MOS and "amplifies" this signal to provide the higher current levels required of the digit drivers. The only components located on the reverse side are the quartz, the trimmer for the quartz, and the two batteries.

A completely different approach to the LED modules is exhibited in a module manufactured by Texas Instruments, shown in Figure 6-5. In this module, the display is preassembled and has a magnifier system molded on to the display package. The integrated circuit (of the I^2L variety) is also prepackaged and mounted on a circuit board. This circuit takes advantage of the properties of bypolar I^2L and has both the segment drivers and the digit drivers manufactured as a part of the timekeeping integrated circuit. Figure 6-6 is a photograph of the

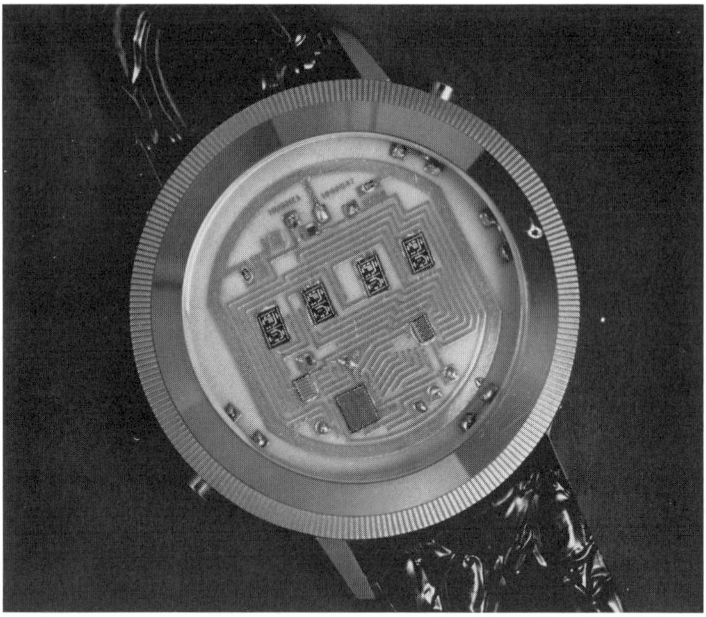

Figure 6-4. Front view of Hughes Aircraft watch module. (*Courtesy* of Hughes Aircraft Company.)

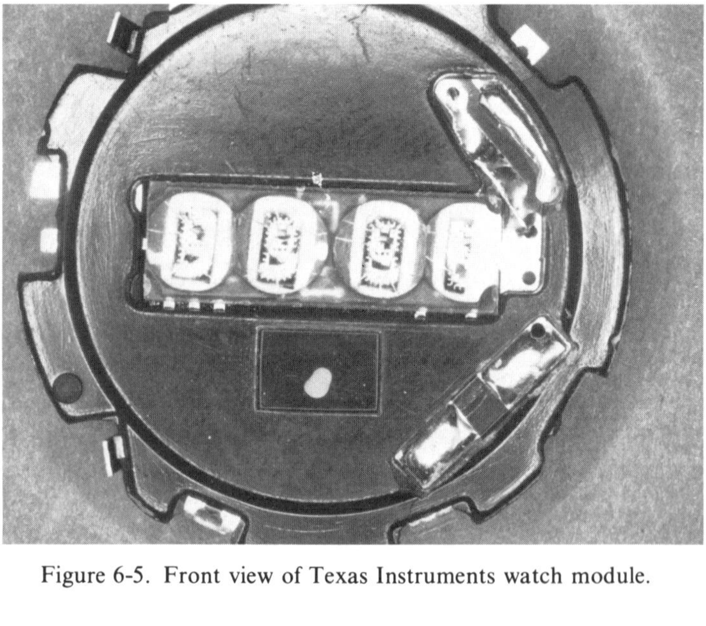

Figure 6-5. Front view of Texas Instruments watch module.

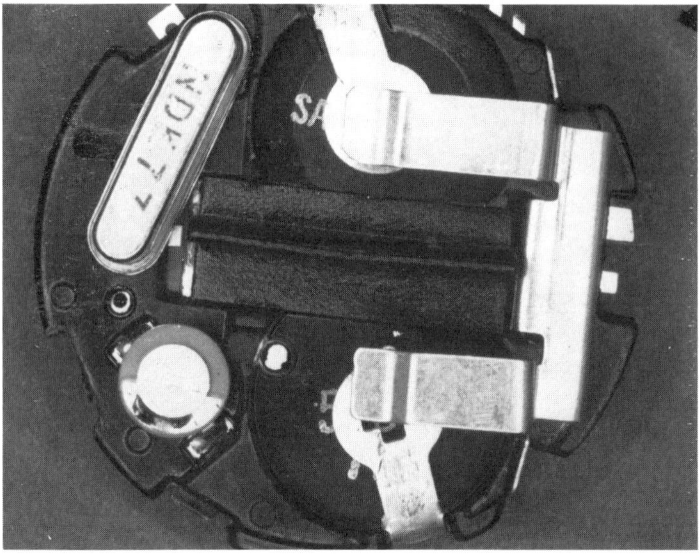

Figure 6-6. Back view of Texas Instruments watch module.

reverse side of the module, showing the two batteries, the quartz and the trimmer capacitor.

NEW DIRECTIONS FOR THE LED WATCH

LED displays are a mature product. For example, where there is need for a display that will show both alphabet and numbers, it is available in the LED, but must often be developed in competing technologies. Therefore, it was to be expected that the first calculator watch, the first alphanumeric watch, and many other firsts were all LED display products.

With the space, power supply, and electronics required for driving five or six alphanumeric digits available in a LED watch, the next step of being able to display a stored message was a natural evolutionary step, and was developed by Hughes Aircraft Company. This has five digits of alphanumeric LED display, segment and digit drivers (as discussed previously), and integrated circuit which computes both time and stores the message. All watches built as of the date of publication of this book are programmed by removing the module

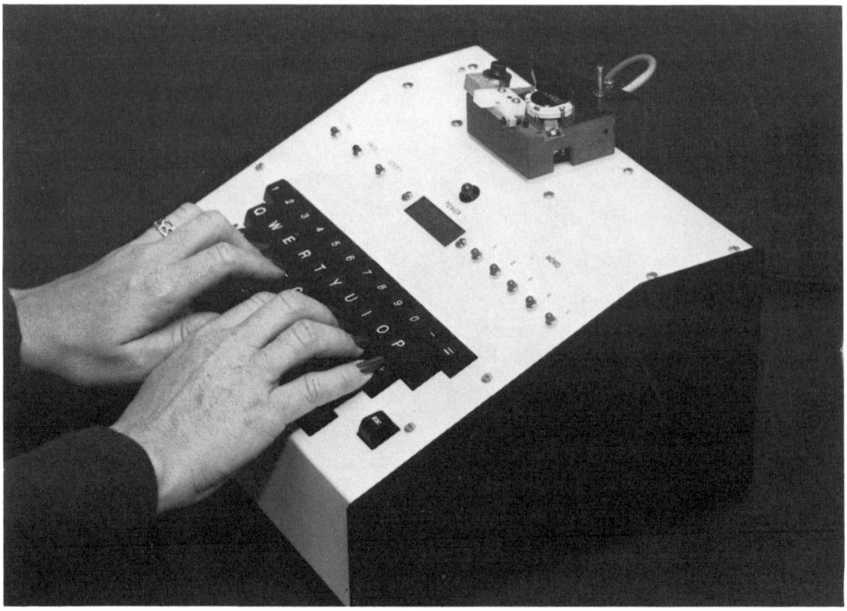

Figure 6-7. Message watch programming console. (*Courtesy* of Hughes Aircraft Company.)

from the case and putting it into the programming console, as shown in Figure 6-7. The words to be entered are simply typed on the keyboard and then with the press of a button, are entered into the module memory bank.

A newer version of this module will be introduced in 1978, which is programmed through the viewing screen by use of infrared light and does not require removing the module from the case. The result of all of this electronics is a module which can show a message, such as the one shown earlier in Figure 5-9.

THE CALCULATOR WATCH

A very dramatic product introduction was made by Pulsar just prior to Christmas, 1975. This was the first calculator watch. Since that time, there have been several others introduced, but the calculator watch

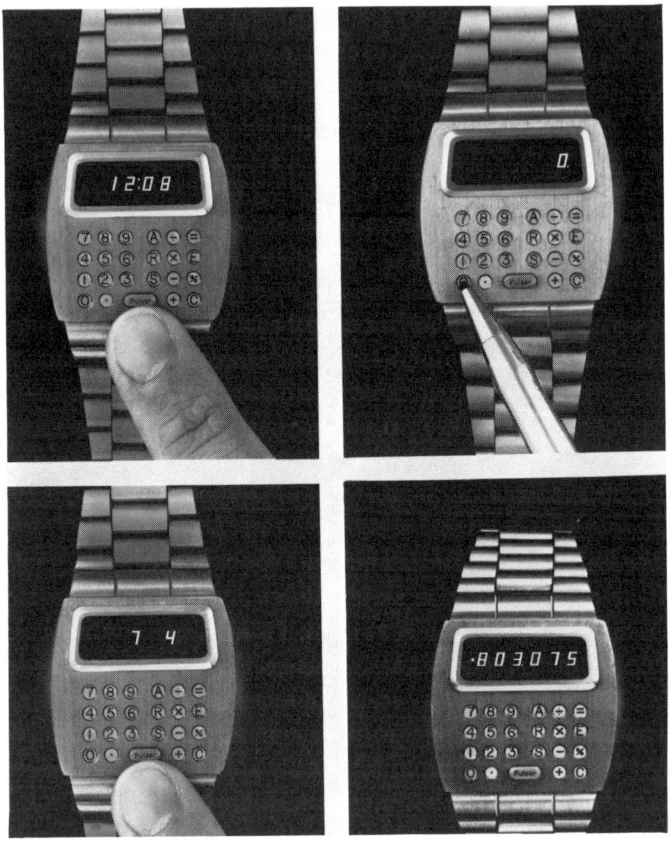

Figure 6-8. Calculator watch operation. (*Courtesy* of Time Computer, Inc.)

manufactured by Pulsar is representative of the type of construction and the type of electronics involved in a calculator watch. The operation of the watch is shown in Figure 6-8. Time is displayed by pressing the button marked "Pulsar." The month and date are exhibited by pressing the button twice in a two second time period. Using either the pen included with the watch or a ballpoint pen, the 0 key can be touched, which converts the watch into a calculator. Calculations are performed and the answer is displayed up to six digits. Calculations up to 12 digits can be performed (only the six most significant are displayed) and full add, subtract, multiply and divide

Figure 6-9. The front of the calculator watch module. (*Courtesy* of Time Computer, Inc.)

features. Memory is also available in add and subtract. There is also a percentage key useful in calculating discounts and markups.

The ability to package all of these electronic functions, the display, and the power source, in a package of only 7½ cc, would have been thought impossible as late as 1970. A photo of the front of the module is shown in Figure 6-9. The six LED display chips are clearly visible across the top of the module, again bonded for scanned or multiplexed operation. A careful examination of this photo also shows some of the details of the operation of the keyboard. In the area of the keyboard, the top layer of plastic has a metallized pattern on the back (or downside). The layer directly beneath this top layer is a spacing layer of the same thin plastic material with a hole pattern punched in the area of the key contacts. All this matches with key contact pads on the main printed circuit board (this is the level that the LED displays are bonded to). The act of pressing the key deforms the top plastic layer

Figure 6-10. The back of the PC board of calculator watch module.

and allows the metallized pattern on this layer to make contact with the appropriate pad on the printed circuit board. As a service note, when the module is removed from the case, the top layer of plastic can be lifted for cleaning the contacts with a solvent and cotton swab.

The backside of the main printed circuit board that makes up the calculator watch module is shown in Figure 6-10. In this photograph, the calculator IC is shown near the right hand edge of the PC board, and the timekeeping IC is shown down near the lower left hand corner. Both of these feed into an integrated circuit that decodes the input information and drives the LED display.

THE FUTURE OF LED WATCH

The principal advantage of the LED is that it uses nothing but mature technologies all the way through—from beginning to end. The IC and the display quite often are manufactured by the same semiconductor manufacturer, and they are compatible with the various means of automation and mechanization chosen by that semiconductor manufacturer. There will continue to be a variety of exploratory products first produced in LED form, but the real advantage of the LED watch is low cost.

The four function calculator has come down from over $100 to its present pricing—being offered very broadly between $5 and $10. It is clear that similar manufacturing economies are available for the plastic cased LED. The LED watch has now and will continue to have the lowest manufacturing cost of all digital watches. As a result of the low manufacturing costs can come the fashion watches for women, the evening watches for men, and knock-about watches for chidren.

7 The Liquid Crystal Display

The Liquid Crystal Display (LCD) is considered by most people as being one type of display, but it is a family of significantly different devices, two of which have been broadly used in watches.

The LED displays discussed earlier were a very fast evolving technology. In just five years, development moved from laboratory discovery to mass production. In contrast, liquid crystals have been known for many years. They were discovered in 1888 and their first use in controlling the flow of light was described in a British patent of 1936. The present form of liquid crystal displays was first displayed by investigators at RCA in 1967 (dynamic scattering) and at Kent State University in 1971 (field effect).

LCDs are fairly common in the watch industry, but the nature of liquid crystal material is still a mystery to many people. Liquid crystals are, as the name implies, materials having properties between those of liquids and those of solid crystals. These properties are generally optical; that is, the ability to in some way interact with a beam of light. It is important to remember that this interaction can be of many different types; there can be direct scattering of the light, modification of polarization, changing the color of the bulk liquid, or color sensitive reflection from the surface. These and others are all part of the liquid crystal effect, and most of them are useful as practical displays. Of these display effects, only two have been widely used in the digital electronic watches.

DYNAMIC SCATTERING LCDs

The first displays to be used in digital electronic watches were of the type called dynamic scattering. These watches had the appearance of characters being formed of a milky substance over a mirrored background. A photograph of this type of watch is shown in Figure 7-1. The construction of such a liquid crystal panel is shown in Figure 7-2. It consists of two pieces of glass, spaced by less than 1/1000 of an inch (0.025 millimeters). On the interior surface of each piece of glass is a transparent but electrically conductive electrode that allows us to apply a voltage to the liquid contained within the cell. The back surface is mirrored to provide the best visibility. When there is no voltage applied to the cell (or in the areas of the cell not having conductive electrodes), the liquid appears to be extremely clear and only the mirrored back surface is seen. With the application of a voltage, a turbulance is created in the liquid which scatters light. The contrast is therefore between the smoothly mirrored part of the surface's background and the frosty appearing areas forming the characters.

Figure 7-1. Dynamic scattering watch.

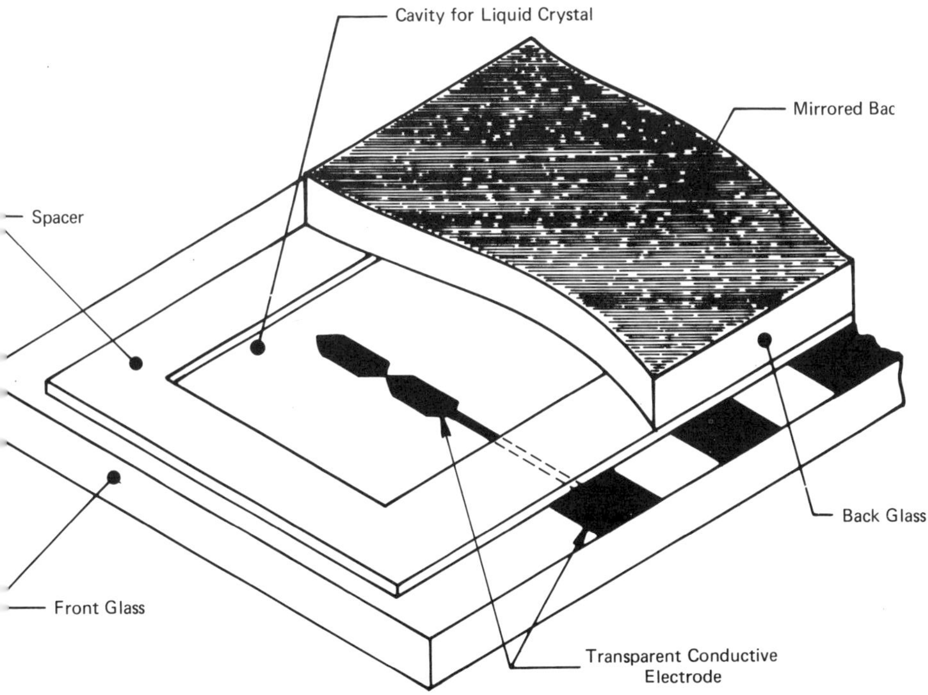

Figure 7-2. Liquid crystal panel construction.

The resulting display can be seen in light levels lower than any other nonlight generating display. It is even possible to see the time in locations as dark as a movie theater. Visibility in such places is not possible with other types of nonilluminated liquid crystal displays.

Dynamic scattering has its best visibility when there is one single light source in a dark room. This is the lighting used for Figure 7-3, and it gives the appearance of white figures against a black background, and a display quality and brightness that is possibly unsurpassed. However, under a different lighting condition, the very same display would become that shown in Figure 7-4. This lighting simulates that of a light colored room with very diffused lighting, or outdoors on a completely overcast sunless day. There is absolutely no contrast between character and background, and no simple way to obtain it.

Some electrical characteristics of dynamic scattering also caused problems. The function of scattering required a reasonably high volt-

Figure 7-3. Dynamic scattering watch spotlighted in a dark room.

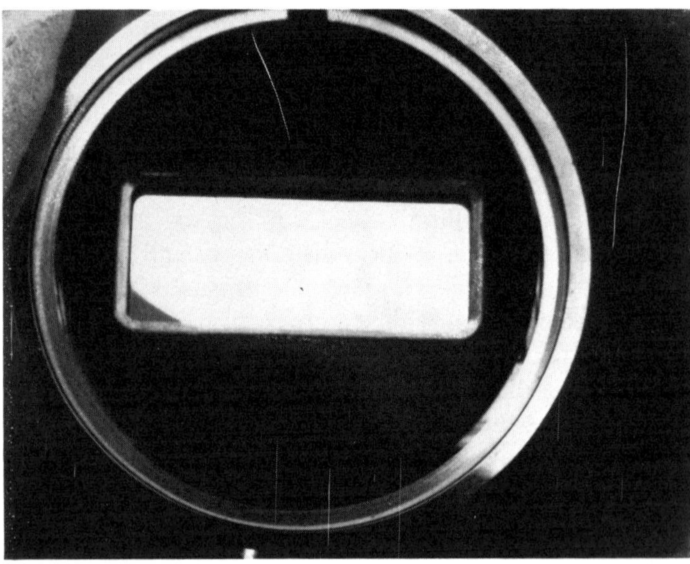

Figure 7-4. Dynamic scattering watch illuminated with diffused light source.

Figure 7-5. Two battery dynamic scattering module.

age, between 10 and 15 volts, and a much higher current than field effect, generally several microamps. The problem presented in a digital watch is that a transformer or coil type voltage converter is required. The power consumption of the early dynamic scattering LCD watches resulted in the use of two batteries for one year's operation. An early LCD module is shown in Figure 7-5 with a transformer and two batteries. This increased module size because of the large battery capacity and space used by the power supply. Clearly, there was need for improvement in visibility, power consumption and the ability to backlight. The field effect display provided this capability.

FIELD EFFECT LCDs

A discussion of the polarization of light is in order to understand the action of field effect displays. We are all familiar with polarized sun glasses. Two lenses from a pair of polarized sun glasses placed against each other transmit full light when aligned in similar fashion, as shown in Figure 7-6. However, when one of the lenses is turned perpendicular to the other, as shown in Figure 7-7, light is not transmitted where the lenses overlap. Now to turn this effect into a watch display.

If there were an object shaped like the letter "T" that had the ability to change light from one polarization to the other, and this object was

Figure 7-6. Sunglass lenses aligned in the same direction.

Figure 7-7. Sunglass lenses aligned perpendicular.

Figure 7-8. "T" shaped object which results in change of polarization.

placed between the two polarizers of Figure 7-6, the light passing through the object would be twisted in polarization so that the light not stopped by one polarizer would be stopped by the other. This would give a black figure artificially created in the shape of a "T," as illustrated in Figure 7-8. Such an effect is the basis of the field effect display.

The liquid crystal material used in field effect displays can, with application of voltage, change the polarization of light. A typical liquid crystal display cell without polarizers is shown in Figure 7-9. This cell consists of two pieces of glass with a minute space between them that is filled with liquid crystal material. The construction is very similar to that shown for the dynamic scattering display in Figure 7-2, except that the mirror is not used on the back and the liquid and manufacturing procedures are slightly different than those used for dynamic scattering.

In Figure 7-10, the display cell of Figure 7-9 is sandwiched between the two sunglass lenses shown earlier in Figure 7-6 and 7-7. You can see that there is no change of polarization in the area outside that occupied by the liquid. In the area of the liquid, however, there is a change of

Figure 7-9. Liquid crystal cell without polarizers.

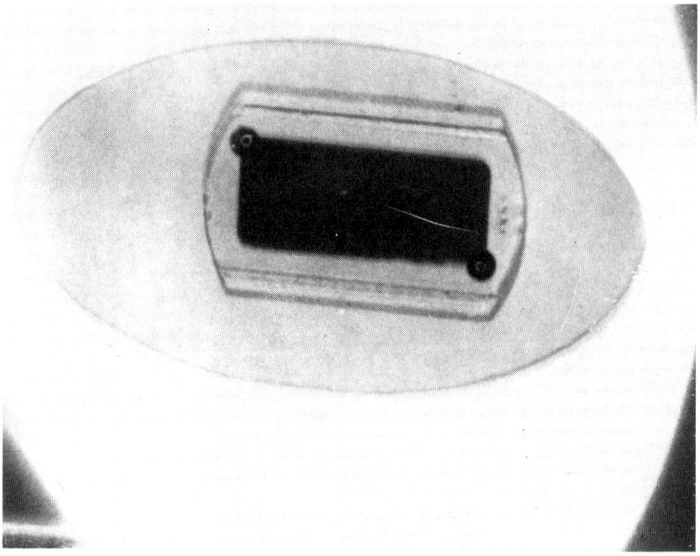

Figure 7-10. Liquid crystal cell between sunglass lenses.

Figure 7-11. LCD with parallel polarizers.

polarization, and therefore no light is transmitted. When voltage is applied to the electrodes, the polarization shift no longer takes place in the area of the electrode. The watch display shown in Figure 7-11 illustrates this effect, with the display actually mounted in a watch module. The display shown has light characters on a dark background. If the front polarizer is changed to be aligned perpendicular to the back polarizer, the display will show dark characters against a light background, as shown in Figure 7-12. This configuration has better low light level visibility than the display as shown in Figure 7-11.

Electrically, the field effect LCD is almost the ideal display. Some manufacturers operate their displays directly from the 1.5 volt battery voltage, while others choose to apply higher voltage from either a voltage doubler or a voltage tripler manufactured as a part of the IC. In general, it requires tighter manufacturing controls for lower voltage, so choice of voltage is an economic trade-off more than a factor of importance for the consumer. Current consumption of the display can theoretically be zero. The action on the liquid crystal material is due to the electric field, hence the name field effect. The current that flows in the display is due either to impurities in the liquid crystal material or to

Figure 7-12. LCD with standard polarizer configuration.

leakages in the display. Normal current consumption is from 0.1 to 0.5 micro-amperes. To put this current into proper perspective, something like 1000 displays could be operated for one year from one single flashlight battery. At this level, the current consumption of the display is a factor of 10 lower than the IC, and therefore even lower current consumption by the display would not significantly benefit battery life.

There is yet another advantage of the field effect liquid crystal display as currently manufactured. The basic liquid crystal cell is completely transparent and has no color. The color of the background and the color of the digits is strictly a function of the polarizers that are used. Polarizers are available in many colors—browns, greens, pinks, blues, magentas, orange are all available. There is the added flexibility resulting from the use of two polarizers. Different color effects can be obtained by using one color polarizer on the back and a different color polarizer on the front. Normally, the visibility of the watches made with the colored polarizers is less than the visibility with black letters on a silver gray background. As liquid crystal displays gain broader use in fashion watches, the availability of the color displays will be of increasing importance.

QUALITY CONSIDERATIONS

It is possible to build a liquid crystal display which in principle will never fail, has a broad viewing angle, good contrast, and other good appearance factors. How far the displays used in the liquid crystal watches deviate from the best that can be made is a measure of the quality of the display.

Liquid crystals are very complex chemicals, and are generally difficult to work with and easy to contaminate. They are in contact with glass and other materials used for sealing the filling hole. Every factor can contribute to the loss of reliability in the liquid crystal display. Many of the early reliability problems were part of the learning stage. The early materials were extremely sensitive to moisture contamination, extremely sensitive to ultraviolet light and comparatively impure. The chemicals used now have relatively low sensitivity to moisture and relatively low sensitivity to ultraviolet light.

At present, the prime sources of failure in displays are mechanical problems; glass breakage, inadequate seal between the front and back glass, or loss of the electrical contact between the front and back glass. To minimize failure due to these factors, the buyer must look to the manufacturer.

The visible determinants of quality are something the buyer himself can judge. Most displays have good contrast appearance when looked at head on. When looking at the display from an angle, several faults begin to become more apparent. Many displays fall off very rapidly in contrast for viewing angles other than head on. Figure 7-13 compares two watch modules that have a similar appearance head on. This photograph is taken 45° off-axis and demonstrates the difference in contrast in displays responsible even at this viewing angle.

A part of the viewing angle problem is that some displays have good viewing angle in one direction and bad viewing angle in another. This effect is referred to as "lobes" and is a result of various manufacturing processes. Several different patterns of the lobing is possible. Limited viewing angles can be at the sides of the display with good viewing from the top and bottom, or viewing can be restricted at the four "corners" of the display.

Along with the limited viewing angle, there is another off-axis problem: the multiple image problem. The numbers look stacked on each

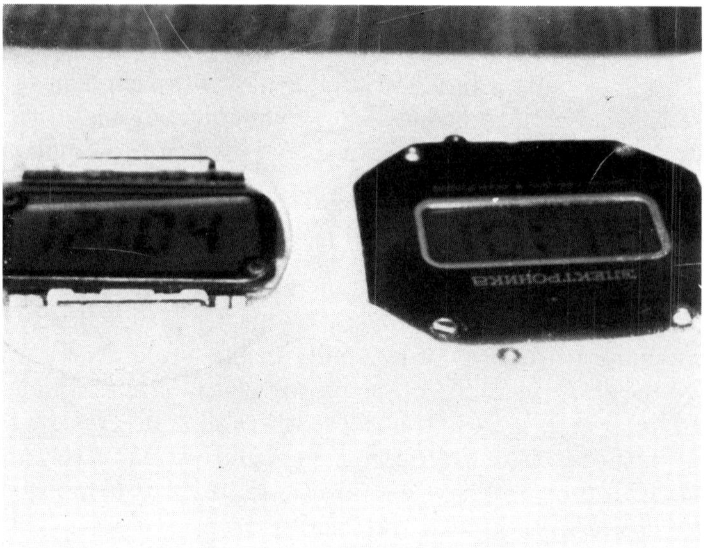

Figure 7-13. Two modules showing good and bad viewing angles.

other like multiple images in a barber shop that has mirrored walls. An example of this type of multiple image is shown in Figure 7-14. Again, the photograph is from the 45° angle.

With the introduction of liquid crystal chronographs, watches with the ability of an electronic stop watch, another defect in liquid crystal displays has become obvious. That is the switching speed. Early liquid crystal displays required several tenths of a second for the character to turn on and off. When the fastest change was one character per second, the character could still clearly be read. Most chronographs now being introduced have a 1/10th of a second function displayed. As a result, display speed has become extremely important. Liquid crystal displays can be manufactured which will change characters in a few hundredths of a second, clearly delineating each digit as it counts from zero through nine in one second. Not all displays are capable of this speed of operation, and on many chronographs the 1/10th second digit is a blur until the stop watch is stopped.

All of the above characteristics combine to contribute to the utility and quality of a liquid crystal display. Each of them becomes important at different times for different applications. The final question is will the display do the job that is needed.

Figure 7-14. Display showing undesirable multiple images when viewed at an angle.

ALTERNATE APPROACHES TO LIQUID CRYSTAL DISPLAYS

Dynamic scattering and field effect liquid crystal displays are not the only ones in the chemists' "bag of tricks." Two others are finding some utilization at the present time.

Cholesterics are temperature sensitive liquid crystals that are used in such applications as digital thermometers and "mood rings." They are a very sensitive indicator of temperature and are not expected to be of importance in digital watch displays.

Pleochroic displays are liquid crystal displays where a dye is dissolved in a liquid crystal and the color of the display changes when voltage is applied changing the alignment of the liquid crystal and dye molecules. Some effort is being made to apply these displays to digital electronic watches, but at the present time, the operating voltage and current consumption are significantly higher than the field effect displays, and the appearance is not improved over the field effects that are available. There is also the problem that all known dyes which are suitable for this use fade in sunlight.

In addition to these true liquid crystals, there are two other effects which should be mentioned. These are very similar in construction to liquid crystal displays, but do not involve the use of liquid crystal material.

Electrochromic displays are an electro-chemical effect which involves the deposition (plating) of a material by an electric current flow to form a character, and then applying electric current in the opposite direction to erase (deplate) the number when a change is necessary. These suffer from problems of high current consumption (compared to field effect liquid crystals) and poor long term reliability.

Electrophoretic are very similar to pleochroic displays, only instead of liquid crystals and dyes, the electrophoretic displays use electrically charged particles suspended in a colored liquid that will move and therefore change color when voltage is applied. Displays of this type that have been built to date have required very high drive voltages, and the appearance is not a significant improvement over that available with other liquid crystals.

At the present time, none of these display technologies appear to have the characteristics necessary to replace the field effect liquid crystal as the prime LCD watch display. New materials and processes would be required to make these useful displays.

The field effect liquid crystal display has a combination of the following characteristics that is unrivaled with any other liquid crystal display technology for watch application: 1) Low voltage—the ability of many displays to operate directly from a 1.5 V source. 2) Low current—the current is lower than the current required by even the integrated circuit. 3) The display is of acceptable quality. The problem remaining with liquid crystal displays is nighttime visibility, and in the following chapter, the Liquid Crystal Watch, we will explore the approaches to solving the night visibility problem.

8 The Liquid Crystal Watch

This chapter will discuss the use of liquid crystal display in a digital electronic watch. Due to the present dominance of the field effect display technology, only field effect implementation will be considered. Other watches are quite similar, for example, those using dynamic scattering.

In order to make a watch from a liquid crystal display, we must provide the proper signals and voltage levels to the display. These are generally provided directly by the integrated circuit; and, of course, there must be a means of connecting the display to the integrated circuit. Quite often with the liquid crystal display there is the need for night time illumination. Finally, all of these components (plus the battery) must be assembled into one operating system and put into a watch case.

ELECTRICAL REQUIREMENTS OF THE DISPLAY

The voltage required to drive a typical field effect liquid crystal display was discussed only briefly in the preceding section. To repeat, if a voltage is applied between the transparent electrodes on each side of the liquid crystal material, then the polarization is shifted and a dark character is formed on a light background. A liquid crystal display cell can, in fact, be driven directly from a flashlight battery and digits will appear. If it is operated for a length of time from a flashlight battery, it has been found that the passage of direct current through the liquid crystal material causes a long-term deterioration of the liquid. All of the liquid crystal displays built today, therefore, require an alternating

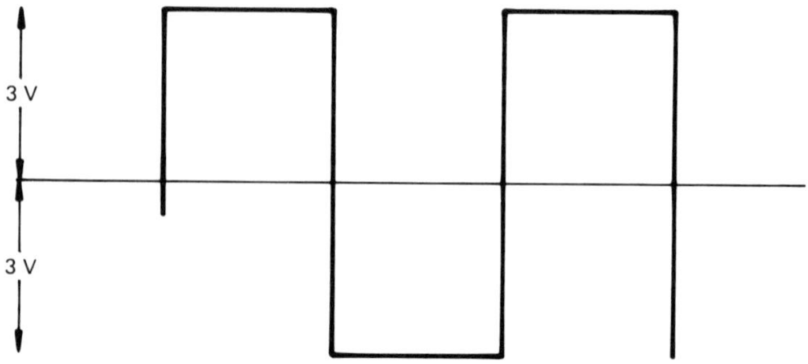

Figure 8-1. Voltage waveform to LCD.

current drive. For a liquid crystal display that is driven at 3 volts, the voltage applied between the electrodes on the two sides of the liquid should look as shown in Figure 8-1. For example, the voltage applied to the front electrode would drive that electrode both plus and minus 3 volts, as compared to the back electrode. In actual practice, both the front electrode and the back electrode are driven to reduce the voltage generated in the watch. As was shown in Figure 8-1, the voltage switches sharply at a rate of 32 hertz. All liquid crystal watches known to the author switch at this rate, and the switching can be picked up by an electrical sensor and used to indicate timing accuracy of the watch.

The current required by each segment (each terminal of the display) is in the order of a few hundredths of a microamp; this is 10-20 billionths of an ampere. Therefore, the outputs of the integrated circuit that drive the display are not built for current capability. The voltage can be seen only using high input impedance oscilloscopes and similar test equipment.

ILLUMINATING THE DISPLAY

One of the early drawbacks in the use of field effect display was its low visibility in dim light, even as compared to the dynamic scattering displays. This night visibility has been solved in two ways. The first solution was the use of a miniature light bulb, powered by the watch battery. Early attempts to use these subminiature light bulbs involved only providing some peripheral lighting for the display itself, and this

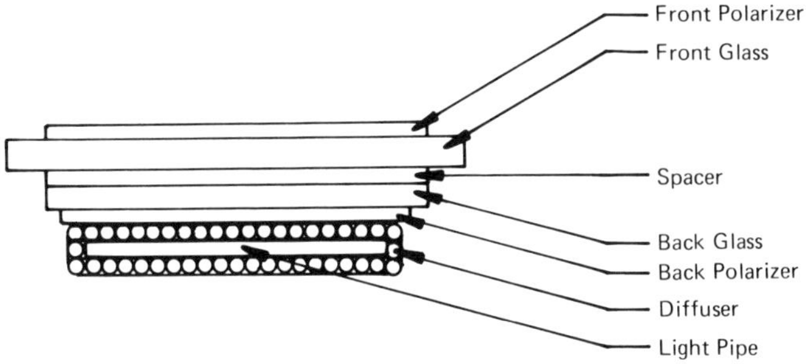

Figure 8-2. Cross section of lighted LCD display.

was generally unsuccessful. The present use of light bulbs requires the use of a "light pipe" across the back of the display to spread the light more or less evenly across the display. A cross section of such a lighted display is shown in Figure 8-2. Notice that the light diffusing panel is located behind the back polarizer of the display. The diffuser panel consists of a light pipe—either of acrylic plastic or glass, which serves to conduct the light from the bulb across the display. Around this light pipe is a diffuser that serves to conduct light out of the light pipe when the light bulb is used. If the light pipe is glass, the diffuser can be a heavy etch. With acrylic plastic, generally a thin foam is used as the diffuser panel. Figure 8-3 shows a typical light bulb-diffuser panel combination used in a modern liquid crystal display watch, photographed in darkness, with illumination only by the light bulb. This shows the relatively even distribution of light across a well-designed diffuser panel.

Recently there has been an alternative to the use of incandescent light—one which doesn't draw any electric power from the battery. This is the use of the radioactive gas, tritium. Tritium is a variation of the very common gas hydrogen. The mechanization is very simple. Light is generated the same way that a television picture is created; electrons given off by the tritium gas strike a phosphor screen that coats the inside of a glass tube. A portion of the energy of the electron is converted into light by the phosphor. A cross section of a tritium tube is shown in Figure 8-4, and a photograph of an actual tube is shown in Figure 8-5.

With any mention of radioactivity, there is always the lingering question: is it safe? This question can best be answered by again turn-

Figure 8-3. Light distribution panel for use in LCD watch.

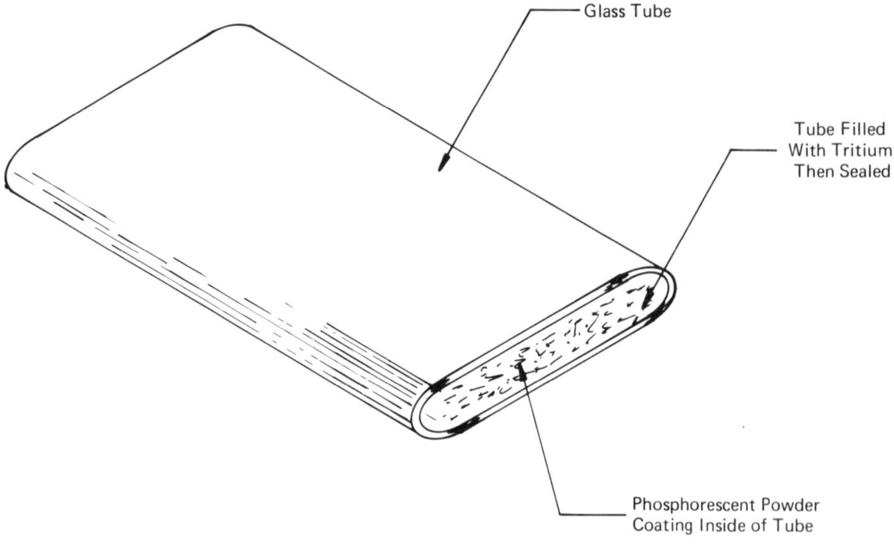

Figure 8-4. Cross section of tritium tube.

Figure 8-5. Tritium tube.

ing to our analogy of the television set. In the television set, electrons are "boiled off" of a heated cathode, then accelerated using a high voltage to a level of 15,000 to 24,000 volts energy level. Electrons in tritium are generated through the decomposition of gas and are sent on their way with a much less penetrating power of 5900 volts energy. These electrons serve only to strike the phosphor (similar to the phosphor used in television tubes). As a result of the low energy of the electrons and the complete glass enclosure, there is no radiation leakage that is detectable with any radiation monitoring equipment.

The total quantity of tritium gas enclosed in the tube is quite small, and even in the unlikely event of breaking 100 tritium tubes in a small closed room, the gas would dissipate and the concentration would not result in medical problems. A properly designed tritium illuminated watch can be easily seen in all lighting conditions, from bright sunlight through dusk to dark, without pressing a button.

MAKING CONNECTION TO THE LIQUID CRYSTAL DISPLAY

Liquid crystal displays are sheets of glass with conductors consisting of thin transparent layers deposited onto the glass. Normal techniques

Figure 8-6. LCD module using spring contact to display.

of interconnection like soldering or bonding, as were used in the LED watch, are not well suited for making connections of this type. The first contacts used on liquid crystal displays were metal springs similar to those that have been used on other glass enclosed displays, such as the plasma displays for calculators. Metal spring contacts were adapted to some of the early liquid crystal display watches, as shown in Figure 8-6. The display was held down onto the spring by a plastic piece that forms the top of the module.

The low current of the liquid crystal display provided the opportunity for newer contact systems—the use of conductive plastics. There was the possibility of using the conductive plastic to achieve better contact over a larger area of the display at lower cost than was available with the springs. A photograph of this type of contact system is shown in Figure 8-7. In this photograph, the white area is a non-conductive plastic and the grey bumps sticking through on both top and bottom are a conductive plastic that is molded into the non-conductive plastic. This system has been made to work, but it required close alignment of both the connector to the printed circuit board and the display to the connector.

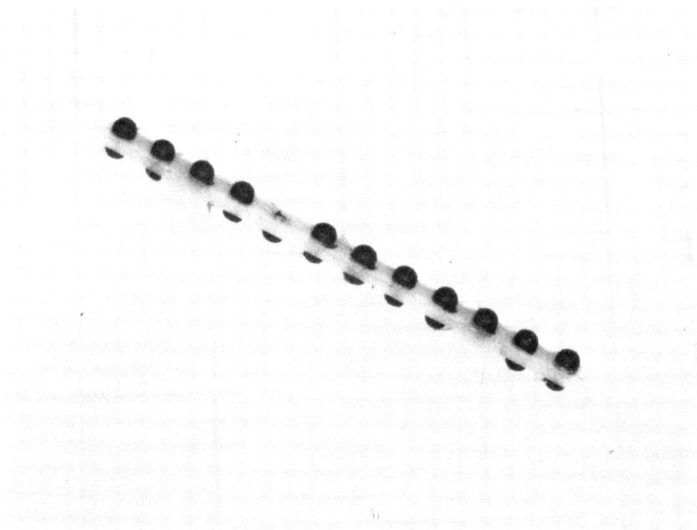

Figure 8-7. Plastic connector with conductive "Bumps."

A simpler lower cost system was devised that reduced the alignment problem, requiring only alignment between the display and the printed circuit board—no alignment involving the connectors. This system involved alternating layers of conductive and nonconductive plastic to form a conductive path between the display and the printed circuit board, as shown in Figure 8-8. With the alternating layers of black conductive plastic and clear nonconductive plastic, it is easy to see how this type of connector was named "Zebra."* A photograph of a Zebra connector in a module is shown in Figure 8-9. This type of connector is the most widely used of all types at the present time.

In earlier paragraphs, it was stated that connection between the integrated circuit and the LCD cannot normally be made by bonding. It is possible, by doing additional processing to the display (firing gold conductors into the top of the usual thin film conductors on the display) that an integrated circuit can be bonded to the liquid crystal display. An integrated circuit mounted by use of this technique directly onto the liquid crystal display is shown in Figure 8-10. In this example,

*Trademark of Technical Wire Products, Inc.

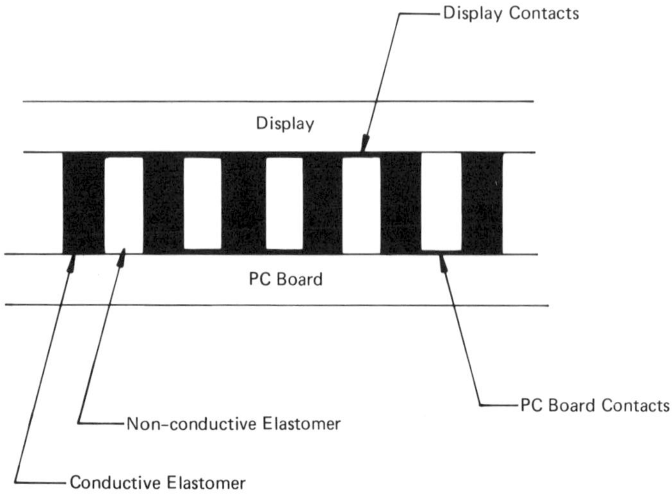

Figure 8-8. Modern plastic connector for use with LCD.

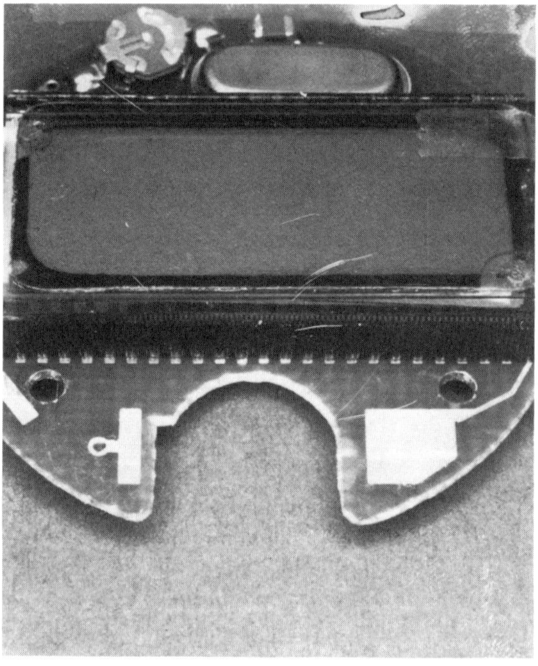

Figure 8-9. LCD module using "Zebra" connector.

Figure 8-10. Display with directly bonded IC.

the IC has been covered with plastic, the leads to the IC from the display are gold fired onto the glass substrate. Contacts from the IC to power and other electrical connections are made by spring loaded pins to the rectangular pad areas shown.

Mounting the integrated circuit directly onto the display results in an extremely compact package which can be very reliable. The problem with this structure is cost. The IC and the liquid crystal display are the two highest cost components of the watch, and if there is either a bad IC or a bad display, the entire assembly must be discarded. As a result, this construction is extremely expensive, and will probably be used only by a small number of manufacturers on premium priced products.

ASSEMBLY INTO A MODULE

Although all modules contain somewhat the same components, the means of putting them together can be completely different. Figure 8-11 shows the printed circuit board assembly of a module for a ladies watch manufactured for the SENSOR and SAVANT watches by Micro Display Systems. This is the side with the integrated circuit covered by the black plastic. The lines of contacts above and below the IC

Figure 8-11. The front of printed circuit board of ladies LCD module by Micro Display Systems.

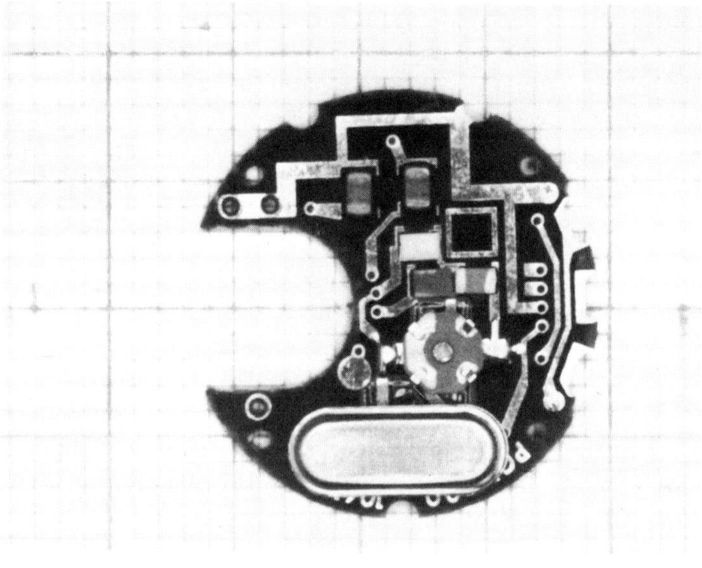

Figure 8-12. The back of a ladies module by Micro Display Systems.

makes contact to the display through the Zebra connector. Figure 8-12 is a photograph of the reverse side of the printed circuit board, showing the quartz, the trimmer capacitor, and the chip component required for the IC operation. An overall view of the completed module is shown in Figure 8-13.

A different approach to module manufacturing is shown by Microma. Their module uses a ceramic substrate instead of a printed circuit board substrate. Figure 8-14 shows the front view of the module with the display in place. Figure 8-15 shows the back view of the module. Here, the ceramic onto which all components are mounted is clearly visible (it is the white square). Contact to the push buttons are made by a pad soldered to the right hand side of the module. A packaged quartz with a transparent window is shown right below the trimmer capacitor.

Modules manufactured by mechanical watch manufacturers look distinctly different than those manufactured by digital watch manufacturers. An example of this is a module manufactured by Seiko, as shown in Figure 8-16. This is the front view of the module, showing the

Figure 8-13. Complete ladies module by Micro Display Systems.

Figure 8-14. The front of a Microma module.

Figure 8-15. The back of a Microma module.

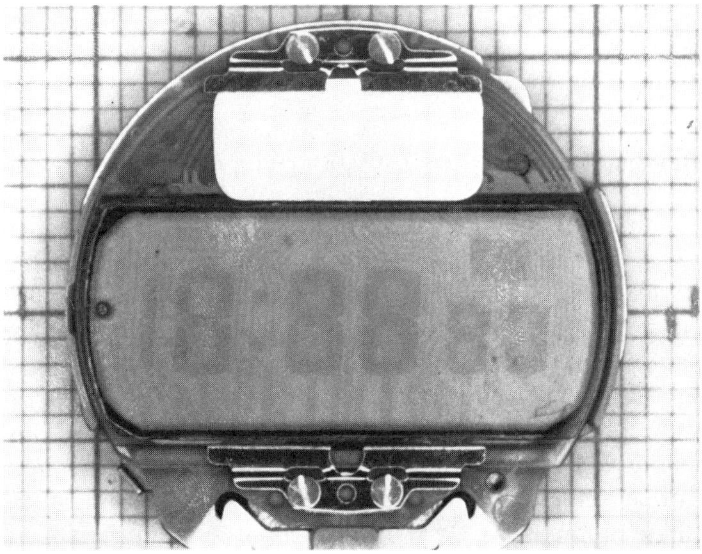

Figure 8-16. The front of a Seiko module.

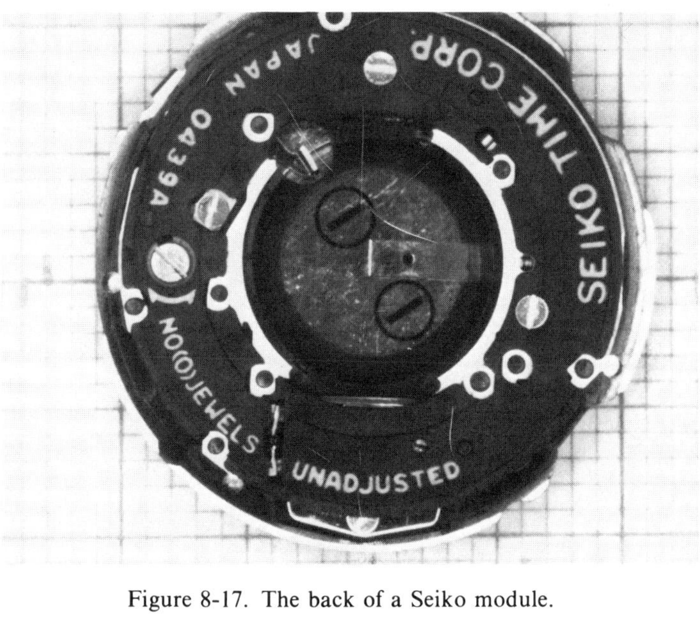

Figure 8-17. The back of a Seiko module.

display held onto the metal frame of the module by two screws upper and two screws lower. Figure 8-17 is the reverse side of the module, showing only the location for the battery and the access to the trimmer.

LIQUID CRYSTAL DISPLAY WATCH TRENDS

There is no doubt that great strides have been made toward solving the numerous early problems of liquid crystal display watches, including display reliability and reliability of contact between the integrated circuit and the display. There is still the limitation on display complexity that results from the need of one contact per display segment. Even this problem may be solved through the recent advances made in multiplexed displays. In this environment, we are seeing the LCD watches establishing a solid position in the marketplace in the middle and upper price ranges. The manufacturing costs will continue to be more than the manufacturing costs of LED modules, and there is no question that mechanical watches will continue to have styling advantages over any of the digital timepieces. Within these areas of constraint, however, the liquid crystal watch is still in a rapidly expanding market.

9

Servicing the Digital Electronic Watch

Many people have traveled the path of rapidly changing technology. The village blacksmith who was asked to repair his first automobile, the radio service technician being brought his first television set, the television technician that understood tubes and was brought his first solid state television set. There are two things that the watchmaker looking at a digital electronic watch must accept. The first is that he is not confident in his ability to diagnose the problem and make the necessary repairs; second, he will make errors—probably many errors—in gaining that confidence.

COURAGE

To the experienced watchmaker, the digital electronic module may appear alien. The purpose of this section is to explain what is known (and not known) that would be of benefit to the watchmaker in understanding these watches.

Backs and bracelets of digital electronic watches are attached as they are in mechanical watches. Bracelets are generally put on with spring bars, and backs are either screwed or snapped. They can be removed with a back knife or a back wrench, as appropriate.

With the back of the watch exposed, this is your opportunity to identify some of the components discussed in earlier chapters of this book. The LED shown in Figure 9-1 clearly shows the location of the two batteries and the tubular quartz crystal. The small rubber plug, which covers the access to the trimmer, can simply be removed with tweezers for timekeeping adjustments on the watch. It is especially

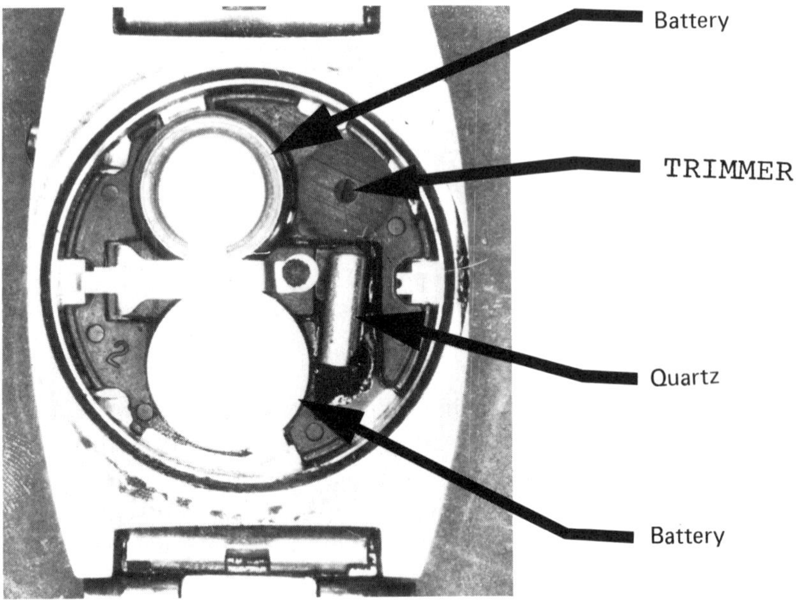

Figure 9-1. Back view of LED watch.

important to note the orientation of the batteries. The battery toward the top of the photograph is mounted with its small end outward, and the battery below it is mounted with its small end inserted first. It is important to maintain proper orientation of the batteries anytime battery replacement is made.

According to one LED manufacturer, 56% of the return for repairs involve replacement of the batteries as the only repair. Therefore, the first repair attempt should be removing the batteries. Before inserting new batteries, inspect the contacts and make sure they are clean and properly shaped for contacting the battery. When inserting new batteries, make sure that they are inserted in the proper orientation!

Some electronic watches require that the case back be on to complete the electrical circuit, so loosely place the case back on the back of the watch and test for operation. If the watch operates, you've got it made (and it will in 56% of the cases). You can then replace the back and present it to a satisfied customer.

Prior to discussing what happens if battery replacement does not solve the problem, let us also examine the open back on a liquid crystal

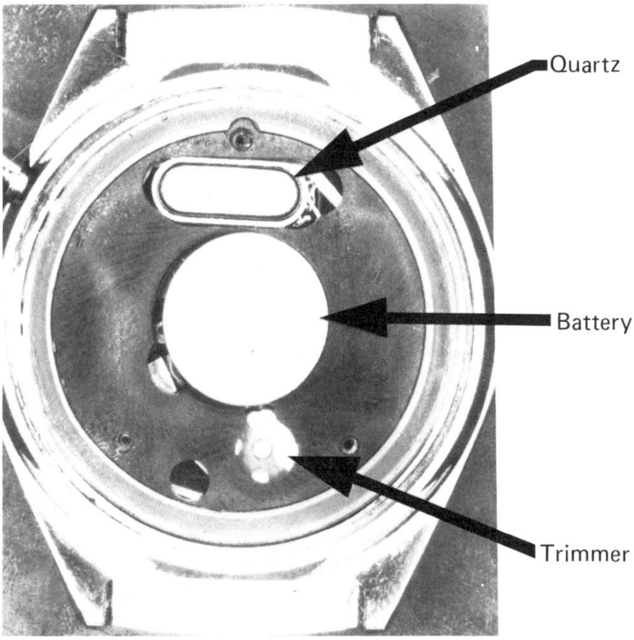

Figure 9-2. Back view of LCD watch.

display watch, shown in Figure 9-2. Again, we see the center location of the battery, the quartz located above the battery, and the trimmer located opposite from the quartz, or just below the battery.

Many of the watches that are returned to the manufacturer following battery replacement are simply in need of being "set up." When a battery is completely run down or when the battery is removed from the watch, all power is moved from the integrated circuit. When a new battery is inserted in the watch, the integrated circuit "turns on," and not necessarily in ways that make either sense or complete characters. It is also possible for the display to remain blank upon insertion of a new battery. All of these conditions can be fixed by going into the set sequence of the watch, as described in the Appendix on Setting and Operating Instructions. This allows the IC to "purify" itself and get rid of nonsensical numbers that might occur when power is applied.

If none of these techniques puts the watch back into satisfactory operating condition, then what to do next? There must be enough information available for an analysis of the future action.

JUDGMENT

If the watch doesn't work with new batteries and clean battery contacts, what do you do then? The best answer to this question comes from looking at the reasons, as collected by various watch manufacturers. These figures do not include case and bracelet problems, nor do they include the battery problems. Let us first consider the failure of LED watches. Numbers have been obtained from one manufacturer that uses nothing but packaged displays and packaged ICs, and another manufacturer who only uses chip components. This information in no way compares the reliability of the two different assembly methods, but only the frequency of failure for various types of problems.

Table 9-1 gives the failure rate for modules built with chip ICs and chip LEDs. The largest cause of problems was the loss of a bonding wire to the IC or bad battery contacts that resulted in the complete nonoperation of the module. Next was loss of contact to one or more segments of the display, generally due to loss of a bonding wire to that segment. A lesser problem was a loss of contact to the various push buttons that made it impossible to call the date or put it into the set mode. The "Other" classification includes cracked or broken substrates, and bad quartz.

Table 9-2 gives a comparable list of problems for a module that uses a packaged display and a packaged IC. Some failure in the display accounts for over half the problems of this type of LED watch. Failures involving switches and contacts is the next in line, and failures involving quartz and the IC are an almost insignificant 3%. The "Miscellaneous" classification includes such items as bad contacts, bad printed circuit boards, and other similar problems.

Looking down this list, it is obvious that to make a significant amount of repair on the module, a repairman would need either semiconductor bonding equipment or replacement displays for the modules.

The situation is similar with LCD watches, as shown in Table 9-3. The most failure prone component is the light bulb which illuminates the display at night, causing over one-third of the problems. Next in line are connection problems to the display, followed by quartz problems, and either broken or bad displays. The "Other" classification includes failures in spring contacts and bad mounting of chip capacitors.

Table 9-1. LED Module with Chip IC and LED Displays.

Nonoperating (Contacts or bad bond to IC)	30.4%
Display Problems (Bad bond to display or IC)	27.3%
Setting or Alternate Mode Problems (Bad contacts or bonding to IC)	10.7%
Other (Cracked substrate, bad quartz, etc.)	31.6%
	100%

Table 9-2. LED Module with Packaged IC and Display.

Display Problems (Loss of contact in packaged display)	55%
Switch Problems (Contact to switches, etc.)	16%
Bad IC or Quartz	3%
Other (Opens or shorts in circuit board, etc.)	26%
	100%

Table 9-3. LCD Module Failures.

Replace Light Bulb	35.2%
Bad Contact to Display	20.8%
Replace Quartz	14.5%
Replace Display	8.3%
Other	21.2%
	100%

The displays on liquid crystal display watches are generally replaceable (watches having the IC mounted directly on the display are an exception), therefore, the liquid crystal tends toward being more repairable than the LED watches. A skilled technician could repair most of the group having display connection problems, without even parts replacement in many cases. Going beyond this, however, to replacing the light bulb or the quartz requires extensive disassembly of the module and an unreasonably large parts inventory.

The big question comes up why don't manufacturers make digital modules that are completely repairable. There are two answers to this question, and they both require some explanation. When television set manufacturers first began manufacturing solid state televisions, many of them used sockets for plug-in transistors and integrated circuits, similar to techniques used with vacuum tubes. They found that the failure and repair problems due to the sockets were more of a problem than the failure of the transistors and integrated circuits. In terms of a watch, the construction of a watch would be compromised by the addition of the many connections necessary to make the various parts of a watch easily replaceable.

The second reason is even more subtle than the first. A car engine, for example (and a watch), is a mechanical device and is subject to mechanical wear. You can expect a car engine to wear a certain amount and certain components would have to be replaced every 100,000 miles. The same is true in a mechanical watch. An electronic watch is a different animal. Nothing moves—there are no parts to wear out. Most of the things that fail are "failures waiting to happen," resulting from manufacturing problems, either of the components or of the completed module. There are some exceptions to this, of course, and these include damage to the watch by abuse, plus the fact that certain springs and push buttons on the watch are subject to wear. The result is that the failure of a digital electronic watch module is much more probable during the first year than it is for the next nine years. By the time a digital watch has reached the end of its warranty period, it should be an extremely reliable product.

DISCRETION

Discretion is a developed sense of knowing when to press ahead and when to go into "strategic retreat." In the digital watch repair business, discretion is knowing when to press forward and effect a repair and when to zip it up and return it to the manufacturer. This discretion is developed from doing it wrong a couple of times and thereby developing judgment. Even assuming there is no repair of the module of digital electronic wristwatches, what does this leave for the watchmaker? At the present time, case related problems, battery replacements, and minor timing corrections account for more than 60% of the problems of both the LEDs and the LCDs that were sampled. The watchmaker is completely capable and uniquely equipped to deal with these problems, and of course, the watchmaker is going to make some repair of the modules—those repairs that involve loose contacts, corroded battery springs or dirty push button contacts can be solved with some judicious bending and the use of common solvents. As manufacturers improve obvious reliability problems—the problems on displays and bonding with LEDs, and the problems with light bulbs and display connections on the LCDs—then the battery, timing and case related problems will get to be an even larger percentage of the digital electronic watches brought in for the watchmaker to service.

What about the possibilities of a module replacement program? Some watch or module manufacturers have suggested replacing defective modules and having their dealers stock replacement modules. This program has met with some limited success, but as the number of modules grows to include the multiple time zone, Chronograph, four and six digit, etc., the number of modules that must be stocked becomes very large and impractical to keep these at more than just a few locations. This is compounded by the fact that there is not now, and will not be before many years to come, any standardization of size between brands. So, the answer to the module changeability problem is that if it makes sense for you with your volume of business to stock a few modules from companies that offer a module replacement program, then by all means do so. It will not be an overall solution, though.

There are certain modules that you, the watchmaker, should *never* try to service beyond changing batteries and adjusting timekeeping. These are the LCD modules containing tritium illumination for the displays. All liquid crystal watches containing tritium are marked on the back with the symbol ^3H, which is the chemical symbol for tritium. At the present time, manufacturers of the tritium illuminated module are prohibited from shipping uncased modules by regulations of the federal government. They are also required to manufacture modules that will not come out when the case is opened to replace batteries. It is not that the government believes that handling one watch module represents a significant health hazard, even if the tubes containing the tritium are broken; it is that they desire that any such breakage due to mishandling be done by the manufacturer where it is less likely to occur and reporting procedures are in place. To repeat, it is completely safe to replace the battery and make timekeeping adjustments on tritium watches, but it is strongly recommended not to try and remove the module from the case.

Discretion on the part of the watchmaker and what he tries to do in digital electronic watches does not mean that the watchmaker should ignore this potentially lucrative field. It means that he should not expect to be able to repair as high a percentage right now of the digital watches brought in as he repairs the mechanical watches that are brought in for service. The average equipped watchmaker can now repair over 60% of the digital watches that are brought in. As module reliability is improved, this number will go to 70%, even without the equipment that it requires for module repair. This is not all bad.

10 The Digital Watch Service Bench

REMEMBER DIGITALS ARE DIFFERENT

The watchmaker must come to grips with the basic and fundamental differences between digital and mechanical watches. Two or three strong mental images might put this into proper perspective. An uncased digital watch module can be thrown into a pile of dry sand, even very fine dry sand, without causing the slightest change in operation of the module. Walking across a carpet (and building up a static electricity charge) and then picking up the digital watch module can completely destroy it. The effect will be worse than the effect of sand on a mechanical watch, because the integrated circuit is destroyed, and, therefore, the module can probably not be repaired.

This fundamental difference between digital and mechanical technologies goes right down to the room environment. Very dry atmospheres have always been chosen for mechanical watch factories, as there is less problem with rust. Low humidity generally results in easier generation of static electricity, and this is a great menace to the digital watch.

Each step that is normally used in the cleaning and repair of mechanical watches must be examined in detail before it is applied to the digital electronic watch. Every digital watch manufacturer has had returned modules which have been dunked and cleaned in some watchmaker's ultrasonic cleaner (some LEDs live through this experience, LCDs do not). The whole point is that once the back is off and the module is out of the watch case, forget all of the principles of mechanical watch repair and do it the electronic way.

HOW NOT TO DESTROY THE WATCH (OR KILL THE WATCHMAKER)

Before a digital watch repair can be effected, precautions must be taken that the module being repaired is not destroyed by the lack of knowledge of the servicer. The care in handling a digital watch module is not grossly different than the care you would give a mechanical watch. The digital watch module is not fragile and is made to be handled by peoples fingers without an undue amount of hazard.

The most probable way that you have to permanently damage a digital module is through either static electricity or accidental application of power line voltages to the digital module. The discussion on establishing a service bench will begin with techniques for elimination, or at least minimizing, these dangers.

Static electricity is generated any time that two nonconductive materials are rubbed together. The action of the nonconducting leather on your shoe soles on the nonconducting nylon fiber of the carpet will generate static electricity. When there is high humidity, the static electricity charge is dissipated by electrical conduction of a moisture film on both your shoe and the carpet. The first action that can be taken is to prevent the relative humidity from going too low. When the relative humidity is about 50% and up, very little static electricity is generated. The easiest approach then is to add a power humidifier and put it close to your bench to use in dry months. A humidity indicator located above your bench will give a constant indication of the severity of the problem. Above 50%, there is very little problem with static electricity. In the 30 to 50% range, caution is indicated, but below 30%, everything that can be done to supress static electricity is needed, and even this may not be enough.

Static electricity reduction at the bench is a relatively straightforward procedure. You want your body, the bench, and the watch module to all be at the same voltage, and the easiest way to accomplish this is to have all of them connected by a wire. It makes sense to have this also a part of the program to eliminate the possibility of accidentally applying 110 volts to the watch module. This can be done by a technique called strapping or grounding. The bench should be set up with a soft metal top—a small sheet of copper or copper backed printed circuit board material works fine. There should be a large electrical wire (copper braid works best) attached between this and a suitable ground (a cold water pipe is generally best). There should be a braid ground also from this to any 110 volt equipment used on the bench, such as the watch timer to be discussed later.

One big source of trouble is in 110 volt soldering irons. It is very easy for electrical leakage to develop between the heating element and the tip. A grounded soldering iron such as the Weller Model MP126 soldering pencil with MP101 base unit, as shown in Figure 10-1, is recommended for watch applications.

Even with everything grounded, it is still possible to blow watch modules because there is one ungrounded source of static electricity, and that is you. Prior to touching any electronic watch module, you must be at the same electrical potential as the watch. There is a device made for use in electronic manufacturing consisting of a cuff and a strap made of conductive plastic. This is the Velostat electrically conductive plastic conductive plastic wrist strap manufactured by the Nuclear Products Division of 3M Company, St. Paul, Minnesota. A properly set up digital watch bench is shown in Figure 10-2.

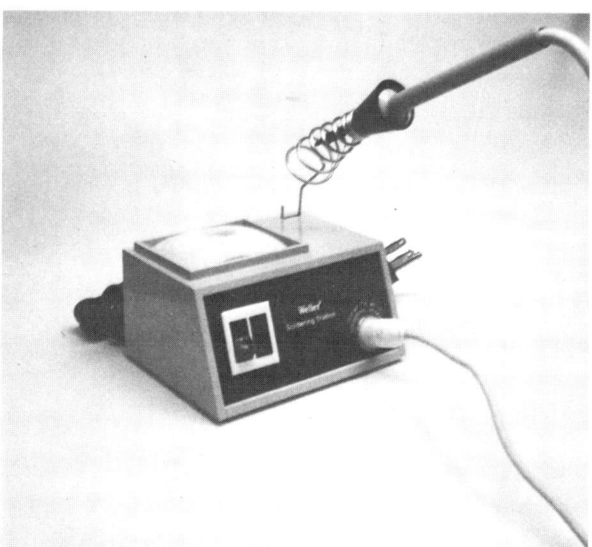

Figure 10-1. Soldering iron suitable for use with electronic digital watch repair.

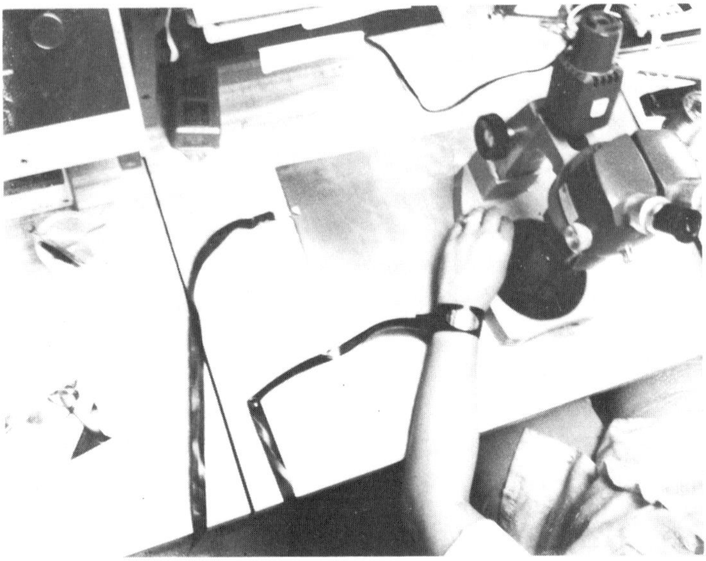

Figure 10-2. Velcro Cuff in use.

A cheaper approach to grounding oneself for static reduction is simply the use of a cliplead from your watch band (if it is metal) to the bench ground system.

A word of caution: operating "grounded" is good for the modules, but if you handle any ungrounded equipment you are asking for trouble. Any defective 110 volt equipment can cause a lethal shock. The controlled conduction of the plastic cuff reduces this risk.

If you, the test and service equipment, and the module are all at the same voltage, the module can be safely handled and a cure effected with a minimum risk.

WHAT EQUIPMENT—WHAT REPAIRS

In the previous chapter, we discussed the servicing statistics of digital modules. By far, the most common problem is bad batteries. A simplistic approach is to plug in new batteries and if the watch works you are an expert. There is much more of a technical impact upon the customer (and upon the watchmaker) to be able to show that a battery

Figure 10-3. Union Carbide watch battery tester. (*Courtesy* of Union Carbide Corporation.)

is, in fact, dead or that it has only a minor amount of life left. For this, a battery tester is needed.

Battery testers for watch batteries are available from $100 down. Special testers are made by watch test equipment manufacturers, by battery manufacturers, and by watchmakers. A specialized watch battery tester made by Union Carbide is shown in Figure 10-3. An adequate battery tester is also offered by Radio Shack at a completely satisfactory price of $7.99.*

The second most common repair problem is bad contacts and connections. Some of these problems are very difficult to repair and require complicated semiconductor equipment—thermal compression bonders and the like. These are beyond the scope of watchmaker repairs. Probably one-half of the contact and connector problems are within the capability of the watchmaker with a minimum of tools.

*Radio Shack and Radio Shack products are referenced due to broad, national distribution. The author believes that comparable merchandise is also available from other electronic supply outlets.

Dirty contacts should be looked at first after replacing the batteries. The battery must make contact, and it cannot connect through corroded contacts. Scraping the battery contacts with a knife and then cleaning them with a swab and solvent such as Freon TF® will fix part of the watches that battery replacement did not.

Beyond replacing batteries and cleaning dirty contacts, the number of modules that the watchmaker can repair is determined by the experience of the watchmaker with that particular module and the indications that the watchmaker might have as to the trouble. For example, if one segment (one part of one digit) of an LCD watch is not functioning, whereas the remainder of the watch is functioning properly, the most probable trouble is a bad contact between the watch module substrate and the liquid crystal display. It is quite feasible for the watchmaker to remove the display, clean the dirty contacts, the Zebra connector (see Chapter 8), and the watch circuit board, and then reassemble to achieve a correctly operating watch. In order to be able to do this the watch repairman must have knowledge and experience in replacing liquid crystal displays. At the present time, most digital watch manufacturers do not provide such instructions. As digital watch manufacturing becomes more mature (and there is less experimentation by the manufacturers) watch module manufacturers will prepare instructions as to how to do those repairs that are considered within the capability of most watchmakers.

In comparison with the LCD problem, if the very same problem occurred with an LED watch (one segment of one number was not illuminated) then the watch would instantly be considered non-repairable, because in all probability it would be either a bad thermal compression bond to the IC or bad thermal compression bond to the display. The watchmaker has no equipment to effect such a repair.

There are other kinds of problems amenable to the repairman. It is not uncommon for a spring or other small part of a digital watch module to actually be loose in the case and fall out when the watchmaker removes the back. This type of problem is generally the result of poor riveting or poor soldering, and a repair very often can be effected even without spare parts available.

Repair of nonoperating modules beyond that discussed must await a maturing of the digital watch technology.

®Trademark of Dupont

THE BIG INVESTMENT—THE QUARTZ TESTER

There is another type of servicing that can be done now, and it can be very profitable for the watchmaker. That is, adjusting the timekeeping ability of the digital electronic watch through use of a quartz timer. This technique is useful not only for servicing, but as a sales aid if you are in the watch selling business. With a quartz watch timer, you can bring the accuracy of quartz watches to better than one second per month, and as the watch grows older and the drift that is natural in the quartz becomes less, it will stay at this accuracy for longer periods of time. The result will be a satisfied customer.

The quartz watch timer is your big investment in quartz watch repair, and it must be chosen with some understanding of what it will do and not do. Also very important is the type of care and service required on the timer itself.

To understand the problem faced with the quartz watch tester, we must first try to understand the difficulty of the job it is trying to do. If the quartz watch is timed within two seconds per month (this is an attainable number) then the quartz crystal at 32,768 hertz must be oscillating accurate to one cycle in 40 seconds. Another way of expressing the frequency stability is parts per million. Two seconds per month error is 0.76 parts per million (ppm). Radio and television stations operate within 10 and 20 ppm of their assigned frequency, some 20 times less accuracy than the watch.

Normal rules of engineering require test equipment to be a minimum factor of 10, preferably a factor of 100, better than the unit that is being tested. A factor of 10 improvement over the accuracy of digital electronic watches places the quartz oscillator in the tester very near to the limits of the best that it is possible to achieve. To be a really effective test instrument, it should be a factor of 100 better than the watch, and this would place the accuracy requirements at a point where it can only be met by atomic frequency standards. As purchaser of a quartz frequency tester, you must be aware that the accuracy of the test box is not significantly better than the accuracy of the watch that you are testing.

In order to make intelligent choices on your big investment, some understanding of the job it does and how it does it is needed. The major portions of the quartz timing box are shown in Figure 10-4. In every timer there is a quartz oscillator, a pick-up device for the watch under

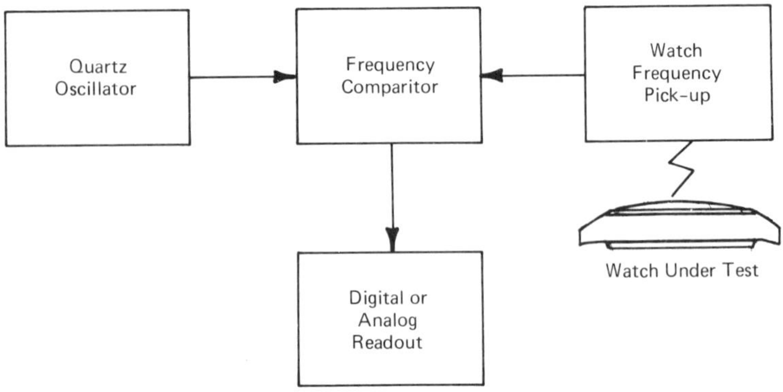

Figure 10-4. Block diagram of quartz watch tester.

test, a frequency comparator, and an analog or digital display of the error, either in seconds per day or seconds per month. The quartz time base is the major contributor to the cost of the timing box. It must be an extremely good frequency standard, having better accuracy than the frequency standards for commercial radio and television stations.

Two types of quartz oscillators are currently in use. One has a high quality quartz in a constant temperature oven. This unit must be kept plugged in and the quartz oscillator portion turned on at all times in order to maintain stability. Another (and less effective) way of compensating for temperature variations is to temperature compensate the quartz. With this type, no effort is made to maintain the quartz at a constant temperature, but a temperature compensation component is put in the oscillator circuit to tend to keep the quartz frequency reasonably independent of the temperature. Without question, quartz frequency standards using constant temperature ovens provide the most accuracy, but satisfactory performance can be obtained from other types of quartz standards, provided they have been properly engineered into the frequency tester.

The next major decision to be made about a quartz watch tester is the device to detect frequency of operation. In the devices used to test mechanical watches, there is a microphone pick-up that is amplified and prints out on a chart of paper referenced to the 60 cycle line frequency. With quartz watches, it is also possible to use a special microphone to pick up the 32 kilohertz quartz frequency for comparison against the standard frequency generated within the tester. Since quartz watches are electronic, there are two other available ways of picking up the quartz frequency. These are electrostatic pick-up (a capacitance probe) or an electromagnetic pick-up to pick up the magnetic fields generated by the flow of electricity within a watch. There are quartz watch testers available that use any one or all three of these types of pick-ups, and they all are serviceable under certain circumstances, and all cause problems under other circumstances.

Analog quartz watches, for example, although outside of the scope of this book, should be able to be tested on the quartz tester. These have metal cases and generally metal dials so the electrostatic or the electromagnetic pick-ups generally do not work well with this type of product. Therefore, an accoustical pick-up is indicated. The LED watch has a large plastic opening with the electronics exposed under the plastic. Either an electrostatic or electromagnetic pick-up can detect the fields generated within the LED watch. (The LED watch must be tested when the display is off! Referring back to the section on quartz, all digital watches change frequency as the battery voltage changes. The load drawn by an LED display from the watch battery causes the voltage to change temporarily during the time the display is showing. Since the display is on such a small amount of the time, something like 30 seconds per day, then the error caused by setting the frequency when the display is not on will be a very small error.)

The liquid crystal watch causes yet another type of problem on quartz timers. These watches quite often have metal masks inside of the case (to take the place of the watch dial), and the liquid crystal display itself can provide some shielding if the back is metalized. Electrostatic, electromagnetic or acoustical pick-up at 32 kilohertz is feasible, but the easiest signal to pick up is the switching of the liquid crystal display itself at a rate of 32 hertz. This frequency is widely used in the design of quartz testers, and is the reason many instruments provide a choice on the front of pick-up frequency, be it 32,768 or 32 hertz.

One popular type of quartz tester is shown in Figure 10-5. This is the Valtec 1000 Quartz Watch Monitor, manufactured by the Valtec Corporation, West Boston, Massachusetts. This tester uses a temperature compensated quartz crystal oscillator as its internal

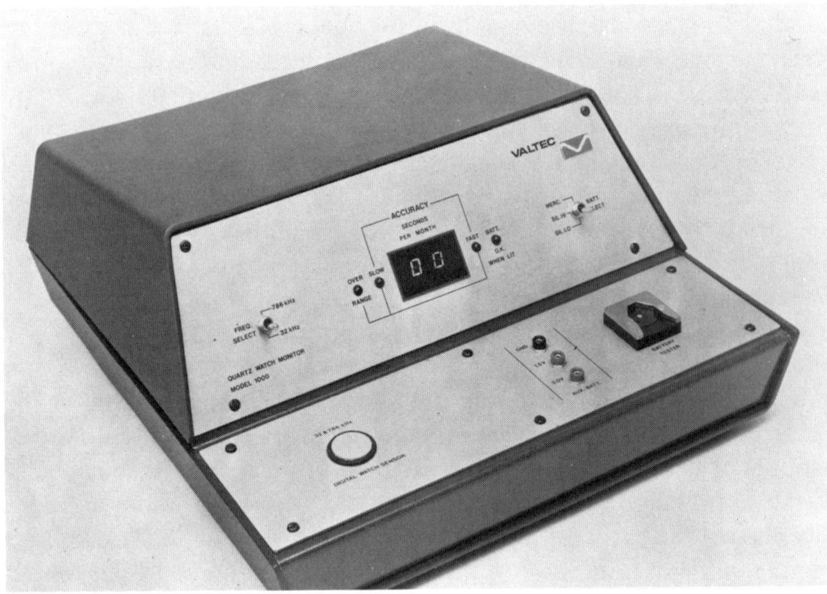

Figure 10-5. Valtec 1000 quartz watch monitor. (*Courtesy* of Valtec Corporation.)

standard, and compares the frequency generated in your quartz watch to the frequency generated by the internal quartz standard. There is a frequency select position so that watches of 32 kilohertz quartz frequency or 786 kilohertz (using modules manufactured by Hughes Aircraft Company) can be measured. The pick-up on the lower shelf, indicated as Digital Watch Sensor, is an electrostatic pick-up of the quartz frequency. This frequency is compared to the internal standard and the accuracy of the watch is displayed in seconds per month either slow or fast. Also as a part of the Valtec 1000 is a battery tester that can be switched to give a proper readout on either mercury batteries or silver batteries, high current for LEDs or lower current as used on LCDs.

With this and all other quartz watch testers, there is a special precaution required. That is, the tester itself must be recalibrated periodically. The quartz oscillator is subject to the same aging, shock and temperature effects described in Chapter 1. The oscillator will change frequency with time, it will change frequency if the box is shocked or otherwise abused. It is necessary that the quartz testers be recalibrated at least once a year, either by the manufacturer or by a local electronic laboratory. This statement is not clearly made on most of the quartz testers examined by the author, but it is nonetheless necessary. If your quartz tester is over one year old, it should be recalibrated. If you are buying one, put a tag on it indicating the date of last calibration and the date of next calibration.

TIME STANDARDS

With a watch capable of maintaining accuracy of a few seconds per month, the use and choice of time standards is very important. The only completely reliable source of time information is provided by the United States and many foreign governments. There are short wave radio transmitters located in Fort Collins, Colorado and in Hawaii that broadcast on 2.5, 5, 10, 15, 20 and 25 megahertz (Hawaii does not broadcast on 20 and 25 megahertz). These stations can be picked up anywhere in the United States on conventional short wave receivers. They are instantly identifiable on the radio with a tick, tick clock beat. Every minute they broadcast the time. The voice announcement in English on station WWV gives the time in Greenwich Mean Time (the time in Greenwich, England). This can be corrected to your local time by adding an appropriate number of hours, but in actual practice, one depends upon local sources like radio stations and telephone time to obtain the proper hour, and upon WWV for precise setting of minutes and seconds.

The easiest way to pick up station WWV is with a receiver made especially for that purpose. One manufactured by Radio Shack, called the Timekube, is a relatively inexpensive receiver that has the choice of receiving WWV or 5, 10, or 15 megahertz.

When setting a watch to synchronize with WWV, date and hours are set to the proper time, and minutes are set ahead to the next minute. In most watches, this stops the watch and it will stay in the locked state until put back into the normal mode when the display button is depressed. Detailed instructions for setting of specified watch brands is given in the Appendix.

Time stations are not just to be found in the United States, but are world wide. Table 10-1 lists the International Shortwave Time Stations, their location and frequency of operation.

Table 10-1. International Shortwave Time Stations.

Call Sign	Place	Carrier Frequency (MHz)
WWV	Fort Collins, Colorado, United States	2.5, 5, 10, 15, 20, 25
WWVH	Maui, Hawaii, United States	2.5, 5, 10
CHU	Ottawa, Canada	3.33, 7.34, 14.67
MSF	Rugby, United Kingdom	2.5, 5, 10
HBN	Neuchatel, Switzerland	5
OMA	Prague, Czechoslovakia	2.5
FFH	Paris, France	2.5
IAM	Rome, Italy	5
IBF	Turin, Italy	5
ZUO	Johannesburg, Rep. of South Africa	10
ZUO	Olifantsfontein, Rep. of South Africa	5
LOL	Buenos Aires, Argentina	5, 10, 15
JJY	Tokyo, Japan	2.5, 5, 10, 15
ATA	New Delhi, India	10
VNG	Lyndhurst, Australia	5.4, 7.5, 12

A CONSERVATIVE POLICY ON REPAIRS

The digital watch industry is still in its infancy—or at least early childhood. Many of the repair problems that are presented to the manufacturers and the repair people are problems of industry youth. These are problems that either have been or will be solved at the next redesign. Acquiring the equipment and skills necessary to make a certain type of repair may well establish you, but in a short lived business. It is a similar problem to that experienced back in the early twenties by people who tooled up to repair wooden spoked wheels for cars, it was just not a good long-term business. Those who wish to speculate on trends in digital watch repair may do so, but for the conservatively managed watch shop, module repairs can be made, but certainly not all module repairs.

Digital watch repair work can be a profitable business. As techniques develop and as manufacturers put out service information, the watch repairman can, with knowledge and confidence in the manufacturer, add the equipment to do those service tasks recommended by the manufacturer. Above all, don't be frightened by—but have fun with—this new technology.

APPENDIX A Manufacturers of Digital Electronic Watches

(With addresses and telephone numbers of Unites States Offices.)

Advance Watch Company
12360 Beech Daly Road
Redford, Michigan 48239
313/937-0800

Alcor Instruments, Inc.
131-27 Fowler Avenue
Flushing, New York 11355
212/445-4000

American Digital
3674 South Bristol
Santa Ana, California 92704
714/556-2912

Bulova Watch Company
Bulova Park
Flushing, New York 11370
212/335-6000

Casio, Inc.
Timepiece Division
15 Gardner Road
Fairfield, New Jersey 07006
201/575-7400

Citizen Watch Co. of America, Inc.
12140 W. Olympic Blvd.
Los Angeles, California 90064
213/826-6541

Commodore Time Co.
901 California Avenue
Palo Alto, California 94304
415/326-4000

Cronus Precision Products, Inc.
2895 Northwestern Parkway
Santa Clara, California 95051
408/988-2500

Croton Time Corporation
Croton Time Building
Croton-on-Hudson, New York 10520
914/271-4737

Data King
291 S. LaCienega - Suite 310
Beverly Hills, California 90211
213/659-0654

Datatime Corporation
756 N. Pastoria Avenue
Sunnyvale, California 94086
408/737-0900

Economic Swiss Time, Inc.
522 Fifth Avenue
New York, New York 10036
212/764-0016

Elgin Watch Company
444 S. Jefferson
Chicago, Illinois 60607
312/454-0828

Fairchild Camera & Instrument
Consumer Products Group
4001 Miranda Avenue
Palo Alto, California 94304
415/493-3100

Gerard Perregaux Corp.
610 Fifth Avenue
New York, New York 10020
212/265-1771

E. Gluck Corporation
417 Fifth Avenue
New York, New York 10016
212/949-5800

Gregtronix Company
6827 Market Street
El Paso, Texas 79915
915/779-6363

Hamilton Watch Company
941 Wheatland Avenue
Lancaster, Pennsylvania 17604
717/394-7161

Helbros Watch, Inc.
2 Park Avenue
New York, New York 10016
212/685-6300

Heuer Time & Electronics
960 S. Springfield Avenue
Springfield, New Jersey 07081
201/467-1890

Hughes Aircraft Company
Solid State Products Div.
2601 Campus Drive
Irvine, California 92715
714/752-6130

Innovative Time Corporation
17875 Sky Park North
Suite G
Irvine, California 92714
714/540-5222

Integrated Display Systems, Inc.
Montgomeryville Industrial Center
Commerce & Enterprise Drive
Montgomeryville, Pennsylvania 18936
215/368-8050

Jupiter Time Corporation
12 West 37th Street
New York, New York 10018
212/594-5877

Leisurecraft Products Ltd.
28 S. Terminal Drive
Plainview, Long Island, New York 11803
516/931-3100

Longines-Wittnauer Watch Co.
145 Huguenot
New Rochelle, New York 10810
914/576-1000

Marcel Watch Company
1180 Avenue of the Americas
New York, New York 10036
212/245-6470

Micro Display Systems, Inc.
4350 Beltwood Parkway South
Dallas, Texas 75240
214/387-0006

Microelectronic Systems Corp.
One Electronics Court
Madison Heights, Michigan 48071
313/399-4300

Microma, Inc.
10260 Bubb Road
Cupertino, California 95014
408/255-8220

Microsonic Corporation
531 Spring Garden Street Mall
Philadelphia, Pennsylvania 19123
215/574-3636

Microstar
1290 Motor Parkway
Hauppauge, New York 11787
516/234-2010

Modulus
2600 Augustine Drive
Santa Clara, California 95050
408/247-4551

Montrex Company
450 W. 33rd Street
New York, New York 10001
212/760-0450

National Semiconductor Corp.
Consumer Products Division
1177 Kern Avenue
Sunnyville, California 94086
408/733-2600

North American Watch Corp.
1345 Ave. of the Americas
New York, New York 10019
212/586-6569

Omega Watch Company
301 E. 57th Street
New York, New York 10022
212/753-3000

Panquartz Digitals of America
7110 North Lyndon
Rosemont, Illinois 60018
312/296-6682

Pulsar
Time Computer, Inc.
P. O. Box 1707
Lancaster, Pennsylvania 17604
717/299-6311

Ragen Precision Instruments, Inc.
77 Water Street
New York, New York 10005
212/344-4050

Pierre Remy
450 E. Middlefield Road
Mountain View, California 94043
415/968-7369

Seiko Time Corporation
640 Fifth Avenue
New York, New York 10019
212/977-2800

Siliconix, Inc.
2201 Laurelwood Road
Santa Clara, California 95054
408/246-8000

Solitron Devices, Inc.
256 Oak Tree Road
Tappan, New York 10983
914/359-5050

Speidel
70 Ship
Providence, Rhode Island 02903
401/421-8600

Suncrux
10440 N. Tantau Avenue
Cupertino, California 95014
408/257-9750

Sundial Enterprises
1170 Broadway
New York, New York 10001
212/679-1188

Texas Instruments, Inc.
13500 N. Central Expressway
Dallas, Texas 75222
214/238-2011

Timelite Corporation
100 Grand
Westbury, New York 11590
516/334-6800

Timex
666 Steamboat Road
Greenwich, Connecticut 06830
203/661-3227

UIT Precision Company
Valley Forge Corporate Center
P. O. Box 801
Valley Forge, Pennsylvania 19482
215/666-9040

Unisonic Products Corporation
1115 Broadway
New York, New York 10010
212/255-5400

Uranus Electronics, Inc.
531 Boston Post Road
Port Chester, New York 10573
914/939-8200

Waltham Watch Company
400 S. Jefferson Street
Chicago, Illinois 60607
312/454-0010

Westclox Division General
Time Corporation
A Talley Industry Company
La Salle, Illinois 61301
815/224-5000

(Western Time Corporation)
Now: RMR Systems
2360-A Walsh Avenue
Santa Clara, California 95050
408/246-1401

Windert Watch Co., Inc.
448 South Hill Street
Los Angeles, California 90013
213/626-7688

Wyler Watch Corporation
315 Park Avenue South
New York, New York 10010
212/777-5002

APPENDIX B
Setting and Battery Replacement Instructions for Digital Watches

The following pages are a collation and systemization of setting instructions for digital electronic watches made by a large number of manufacturers. This information is based upon material provided by the digital watch manufacturers, but limitations to the quality and usefullness of these instructions should be understood by the jeweler, watchmaker, and others servicing electronic watches.

1. Digital watches have gone through various stages of immaturity; every manufacturer has models that they would rather forget. Therefore data is rarely presented on all watches sold by a manufacturer—only on the popular (or successful) ones.
2. There are, occasionally, errors in information provided by the manufacturer. This information itself as presented in this book, is subject to errors.
3. New watches with new and different setting functions are being introduced all the time.

One redeeming value of the information presented is that there are fewer different setting instructions than there are different watches manufactured. There are significant similarities between watches of various manufacturers due to either use of a common module source, a common integrated circuit source, or identical functions by integrated circuits produced by different manufacturers. Each person exposed to a wide variety of digital watches will find that even though his brand may not be listed, the operation and setting instructions can be deduced and referenced for future use.

In the instruction pages, all LED watches are listed first, followed by a brief section of the combination LED/LCD, and finally, a listing of the LCD watches. Within the

LED and LCD sections, the watches are separated by function. If a watch shows hours and minutes only, it is considered two functions. For purposes of this book, a flashing colon, a low battery indicator, or a backlight are not considered as functions. Where there are more than one type of a watch of a given number of functions, they are arbitrarily assigned type numbers for usefullness in cataloging. The watches listed under a given type number *must* only share common function and setting instructions. This grouping does not necessarily imply a common module source or even a common integrated circuit.

Next in the listing are the manufacturers whose watches share that particular function and setting instruction. Beside each manufacturer is the battery recommendation by that manufacturer. This battery (when only one is listed) should clearly be the first choice when a selection is available. When the battery specified is not available, the interchangeability guides provided by the various battery manufacturers can be used with some confidence.

With each watch type there is at least one drawing showing the location of the display button or buttons, and the setting button or buttons. When more than one button location plan is used for a function type, then the number of different button plans are given as needed, and these are keyed to the various manufacturers by letters of the alphabet ("A," "B," "C," etc.). The button plan is presented as it appears looking at the face of the watch. (*Note:* Nothing except the push button locations is meaningful on the drawings—shape of case or window opening is not intended to represent a real watch.)

GENERAL NOTES ON BATTERY REPLACEMENT

Replacing the batteries in a digital electronic watch is the most common maintenance performed by the jeweler or watchmaker. However, it is an area where some care and expertise is needed. If the batteries removed from a watch show signs of electrolyte leakage, or if the battery contacts show any signs of corrosion, the contacts should be cleaned using Freon TF®. If this does not do the job, then the contacts should be further polished with a pencil eraser. If they cannot be restored to a satisfactory condition, the watch should be returned to the manufacturer.

Be sure to insert the batteries properly. (The same drection as they were removed from the watch.) Wrong battery orientation can damage the watch. After installing batteries, hold the back over the back of the watch to complete the electrical circuit, and then check for operation. Assuming all is well, check the "O" ring of the battery hatch or the watch back for damage and lightly lubricate with silicone grease for best seal upon installation of the back or the battery hatch.

Other points of caution; use only plastic (or plastic tipped tweezers) when handling the battery—metal tweezers will short the battery. For other information on battery storage, etc., please refer to Chapter 4.

GENERAL NOTES ON SETTING

Setting most digital electronic watches involves the use of push buttons to do the job normally done by turning the crown on a conventional mechanical watch. On most digital watches, setting occurs through the use of a recessed set button, which must be pushed by something like a ballpoint pen. For the watchmaker or jeweler, a more professional approach is to either purchase a setting tool, or manufacture your own from a plastic rod sharpened to a point in a pencil sharpener.

Most digital watches manufactured at the present time use a sequential setting; that is, the set button is depressed to go from setting one function to setting the next function, etc. At each function, the setting can be changed by pushing the display button. Earlier digital electronic watches had independent setting, where two or more set buttons and the display button were depressed in various combinations to effect the setting. The jeweler or watchmaker will need a special tool to depress two set buttons simultaneously. This special tool can be made most easily by buying a divider with replaceable point at a local drafting supply house. The metal points can be replaced with plastic points sharpened as required.

In each setting instruction, there is the reference to a time tone for setting the digital electronic watch to the precise setting. This time tone occurs each minute at 00 seconds as given by radio stations such as WWV, or others listed in Chapter 10.

The author is indebted to the American Watchmakers Institute for providing setting instructions which were used as a source for most of the following pages.

TWO FUNCTION: Hours, Minutes **LED**

TYPE: 1

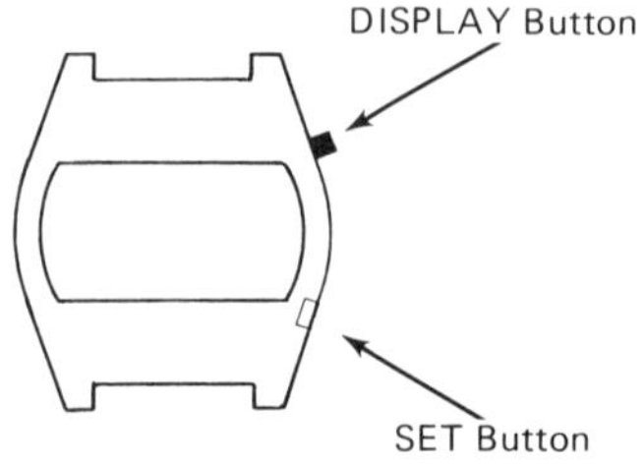

MANUFACTURERS:	RECOMMENDED BATTERY:	OTHERS:
Arnex Quartz	UC386	________
Microsonic	UC386	________
Windert WHL-2F-1	UC386	________
Wittnauer Lady Polara	UC386	________

OPERATION:

To display the time, press the DISPLAY button momentarily and release it. The display will flash on the hour, turn itself off, then show the minute. Each operation takes half a second. Continued pressing of the DISPLAY button will repeat the display sequence after a 1½ second delay.

SETTING INSTRUCTIONS:

Minutes: While holding the DISPLAY button down, press the SET button until the correct minute is shown. Release both buttons. Minutes are now set.

Hours: Set hours by pressing the SET button until the correct hour appears.

The procedure outlined above enables you to set the watch to the nearest minute. If accuracy to the exact second is desired, use the following procedure for setting minutes, which requires telephone or radio time signals:

> First depress the DISPLAY button and continue holding it down throughout this operation. Listen to the next minute time signal. At this instant, press the SET button and hold it down until the correct minute flashes into view. Minutes are now set to the exact second.

TWO FUNCTION: Hours, Minutes **LED**

TYPE: 2

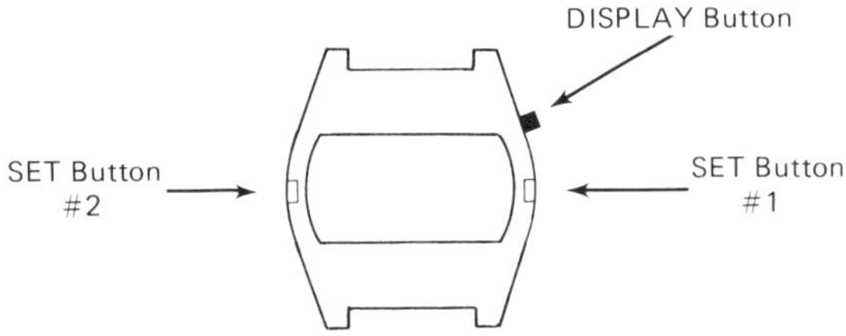

(A)

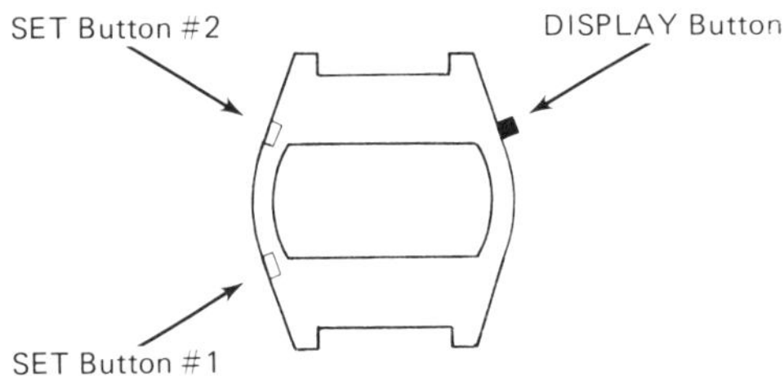

(B)

MANUFACTURERS:	RECOMMENDED BATTERY:	FIG.:	OTHERS:
Avatar	RW44, UC386, M10L124	A	________
Litronix	RW47, UC392	B	________
Windert WL-2	RW44, UC386, M10L124	A	________

OPERATION:

To display time, press the DISPLAY button. The display will remain on as long as the button is depressed.

SETTING INSTRUCTIONS:

Hours: Depress the SET button #1 until the correct hour is displayed. When traveling through different time zones, only this correction is needed.

Minutes: Set tens of minutes by depressing the SET button #2 and the DISPLAY button. Continue holding these buttons depressed until the correct tens of minutes is displayed. Immediately release both buttons.

Set minutes by simultaneously depressing both SET button #1 and SET button #2. Continue depressing both buttons until the correct minute is displayed, and allow the display to then roll to the next minute (in effect the display will now read one minute fast).

Release both SET buttons immediately. Depress SET button #3 and hold it in until the next full minute time signal occurs. Immediately release SET button #3. This action automatically resets SECONDS to zero.

THREE FUNCTION: Hours, Minutes, Seconds **LED**

TYPE: 1

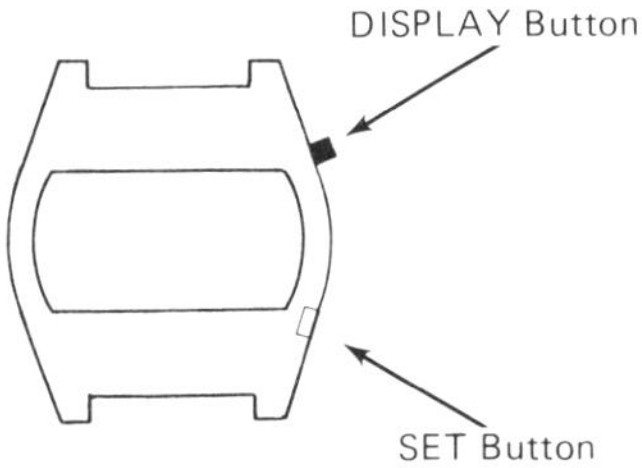

(A)

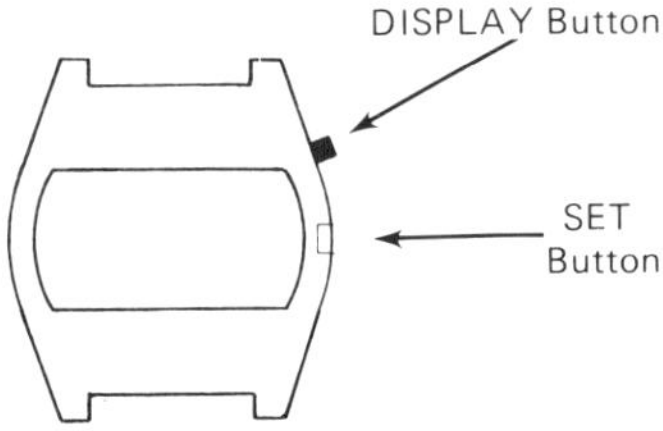

(B)

MANUFACTURERS:	RECOMMENDED BATTERY:	FIG.:	OTHERS:
Alcor	UC357	A	
Exelar	UC357, RW22	A	
Marcel	UC357, RW22	A	
Windert WN-3F	UC357	B	

OPERATION:

Press the DISPLAY button to show hour and minutes. Continue holding the button, and seconds will automatically appear after about 3 seconds.

SETTING INSTRUCTIONS:

Hours: Depress the SET button until the correct hour appears.

Minutes: Depress the DISPLAY and the SET buttons at the same time until the correct minute appears.

Synchronizing the Watch Seconds to a Time Signal: Advance the minutes to the next minute coming up. Continue holding the DISPLAY button. At the time signal, press the SET button too, and immediately *release* both buttons. The seconds are automatically reset at zero and begin counting.

THREE FUNCTION: Hours, Minutes, Seconds **LED**

TYPE: 2

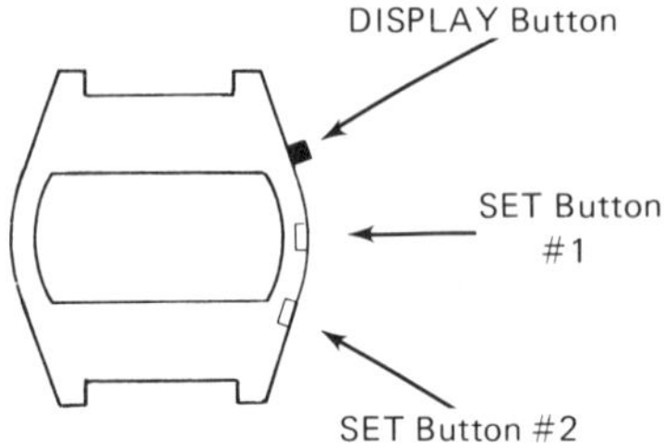

MANUFACTURERS:	RECOMMENDED BATTERY:	OTHERS:
Datatime	UC357	

OPERATION:

Press the DISPLAY button to show hours and minutes. Continue holding the button, and seconds will automatically appear after about 3 seconds.

SETTING INSTRUCTIONS:

Hours: Depress SET button #1 to cause hours to advance. Release button when correct hour is reached.

Minutes: Depress SET button #2 to cause minutes to advance. When setting the minute, the second will automatically advance to zero.

FOUR FUNCTION: Hours, Minutes, Seconds, Date **LED**

TYPE: 1

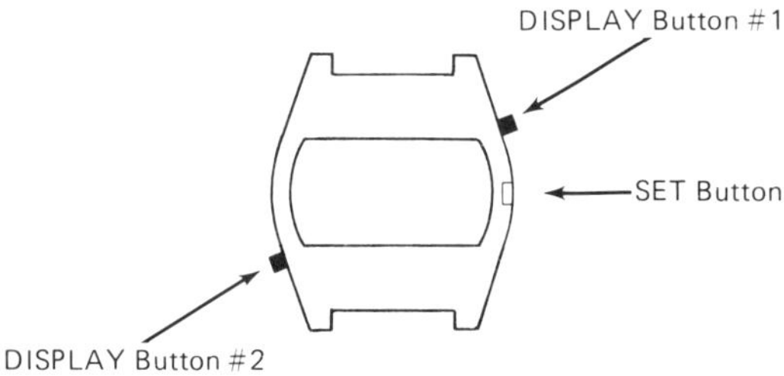

MANUFACTURERS:	**RECOMMENDED BATTERY:**	**OTHERS:**
Arnex Quartz	UC357, RW22	
Benrus Techniquartz	UC357, RW22	
Elgin	UC357, RW22	
Helbros	UC357, RW22	
Longines	UC357, RW22	
Microsonic	UC357, RW22	
Sears Phasar	UC357, RW22	

OPERATION:

To display the time in hours and minutes, press DISPLAY button #1 momentarily. The time will remain visible several seconds after the button is released. To display the date, press DISPLAY button #2. During A.M. hours (midnight to 11:59 A.M.), the A.M. dot (·) will appear whenever the date is displayed. Seconds will appear as a flashing display when both DISPLAY buttons are pressed simultaneously.

SETTING INSTRUCTIONS:

Minutes and Seconds: Depress DISPLAY button #1 and continue holding it down throughout this setting operation. Listen for the next full minute signal; at this instant, press the SET button and hold it down until the correct minute flashes into view. Release both buttons. Minutes and seconds are now set.

Hours: Depress the SET button until the correct hour is shown and the A.M. dot is correct (check the A.M. dot by depressing DISPLAY button #2. The dot should appear only during A.M. hours—midnight to 11:59 A.M. If the A.M. dot is not set correctly, the date will change at noon instead of midnight.)

Date: Set date last. While holding DISPLAY button #2 down, press the SET button until the correct date appears.

FOUR FUNCTION: Hours, Minutes, Seconds, Date **LED**

TYPE: 2

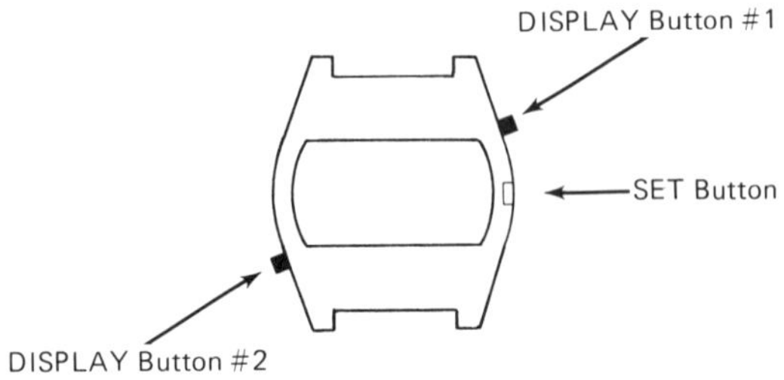

MANUFACTURERS:	**RECOMMENDED BATTERY:**	**OTHERS:**
Arnex Quartz Type M	UC386, RW44	__________
		__________
		__________

OPERATION:

To display hours and minutes, depress DISPLAY button #1 and release. To display date and seconds, depress DISPLAY button #2 and release.

SETTING INSTRUCTIONS:

These Instructions Must Be Followed in Sequence Outlined Below

Date: Press DISPLAY button #2 to display the date and seconds. While pressing DISPLAY button #2, also depress the SET button to advance the date at a 1 Hz rate. Hours, minutes, and seconds continue to keep normal time. When the desired date is displayed, release the SET button.

Hours: Depress DISPLAY button #1 to display the hours and minutes. While pressing DISPLAY button #1, also depress the SET button to advance hours at a 1 Hz rate. (When both DISPLAY button #1 and the SET button are depressed the minutes turn off and an "A" or a "P" is displayed in the right-most digit position to indicate A.M. or P.M.). Minutes and seconds continue to keep normal time but date incrementing is inhibited. When the desired hours are displayed, release the SET button.

Minutes: Press only the SET button to display the hours and minutes, reset seconds to 00, and advance the minutes at a 1 Hz rate. Hours incrementing is inhibited. When the displayed minutes are one minute ahead of actual time, release the SET button. The display turns off, but timekeeping does not begin (seconds are held to 00) until either DISPLAY button #1 or DISPLAY button #2 are pressed, preferably in synchronism with a reference time source.

FOUR FUNCTION: Hours, Minutes, Seconds, Date **LED**

TYPE: 3

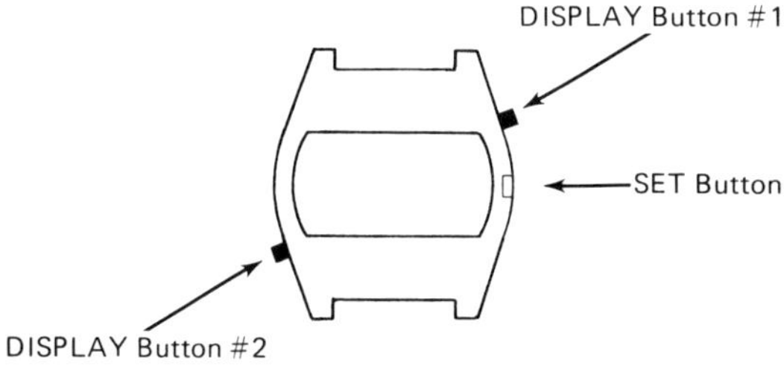

(A)

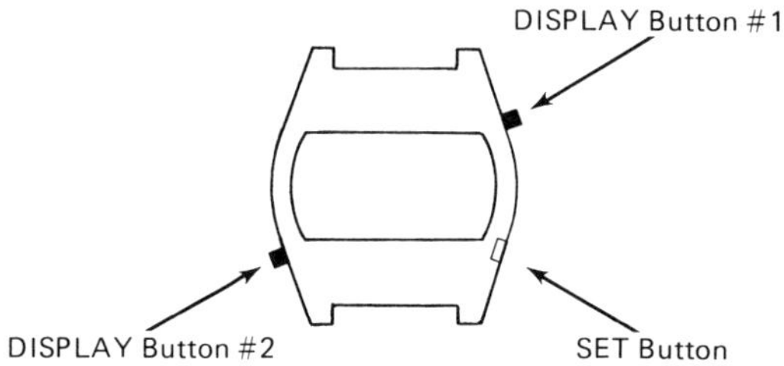

(B)

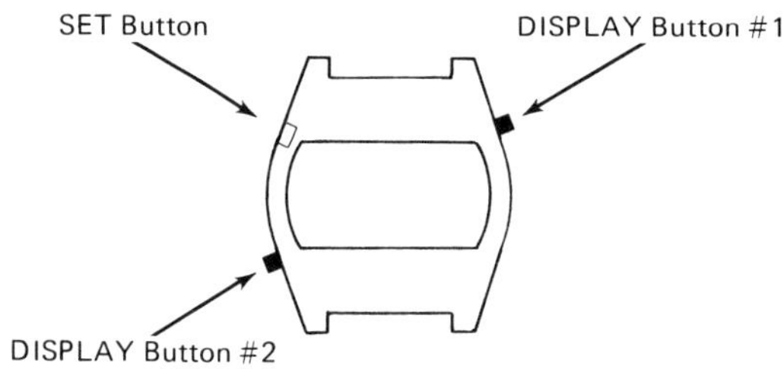

(C)

MANUFACTURERS:	RECOMMENDED BATTERY:	FIG.:
Avatar Series 400	RW44, UC386, M10L124	A
Novus 2 Button	RW22, UC357	B
Saturn	RW11, M10L129	C

OPERATION:

The correct time is displayed by depressing DISPLAY button #1 momentarily. Holding DISPLAY button #1 depressed will cause the display to roll to seconds. Pressing DISPLAY button #2 will cause the display to show the date.

SETTING INSTRUCTIONS:

Minutes: Depress DISPLAY button #2 and SET button simultaneously and hold until correct minutes appear. Release both buttons immediately.

Hours: Depress DISPLAY button #1 and SET button simultaneously and hold until correct hour displays. Release both buttons immediately.

Date: Depress DISPLAY button #1 and DISPLAY button #2 simultaneously and hold until correct date appears. Release both buttons immediately.

FOUR FUNCTION: Hours, Minutes, Seconds, Date **LED**

TYPE: 4

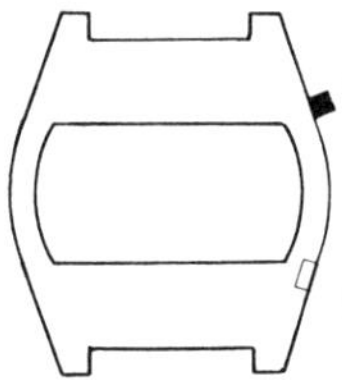

MANUFACTURERS:	**RECOMMENDED BATTERY:**	**OTHERS:**
Novus 1 Button	RW22, 42, UC357, M10L14	__________
Marcel	RW22, 42, UC357, M10L14	__________
		__________

OPERATION:

Depress the DISPLAY button to show hour and minutes. Continue to depress the DISPLAY button, and hours and minutes disappear, and seconds will show. Depress the DISPLAY button twice, and date will appear.

SETTING INSTRUCTIONS:

Minutes: Depress the DISPLAY button and at the same time, depress the SET button. Hold both buttons until correct minutes appear. Seconds automatically reset and will not restart until the DISPLAY button is depressed.

Hours: Depress the SET button only until the correct hour appears. Immediately release the button.

Date: Press the DISPLAY button twice consecutively and immediately press the SET button until the correct date appears. Immediately release the buttons.

Caution: This watch shows two 12 hour cycles. In order for date to change at 12 midnight, it must be set on the proper cycle. To accomplish this, an added feature is a dot in front of the hour, on the A.M. cycle only. If you are resetting before 12 noon, the dot *must* appear. If this dot does not appear, hold the SET button and allow hours to flash. The dot will then appear.

FOUR FUNCTION: Hours, Minutes, Seconds, Date **LED**

TYPE: 5

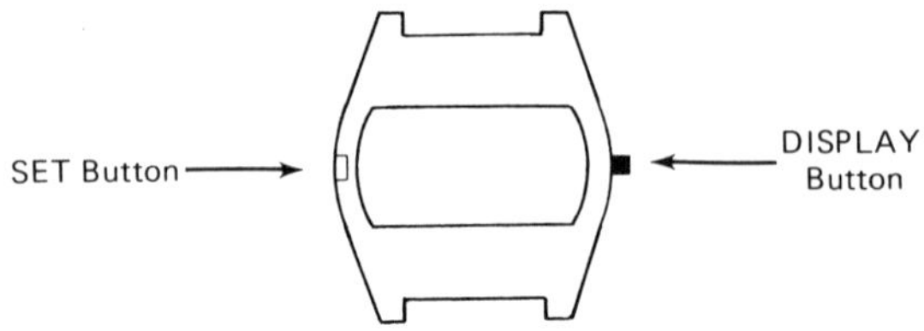

(A)

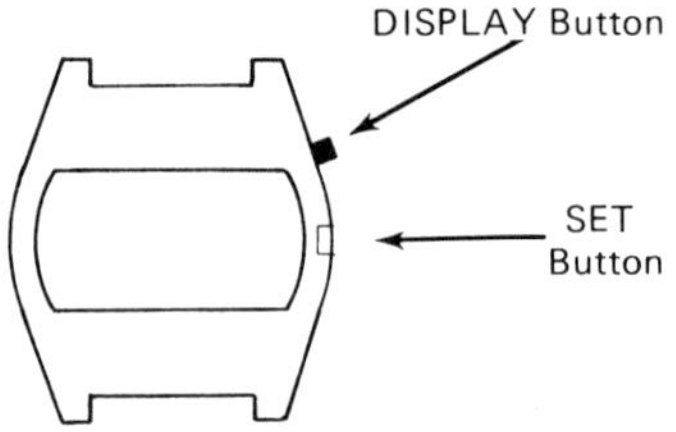

(B)

MANUFACTURERS:	RECOMMENDED BATTERY:	FIG.:	OTHERS:
Girard Perrigaux	UC357	A	________
Wyler	UC357, RW22	B	________
			________

OPERATION:

Depress the DISPLAY button, and hours and minutes will be displayed as long as the button is held in. Note that the color will remain constant during the A.M. and will be blinking when it is P.M. While the display is still illuminated, again depress the DISPLAY button and the date and seconds will now appear.

SETTING INSTRUCTIONS:

Hours and Minutes: While depressing the DISPLAY button to show hours and minutes, simultaneously depress the SET button. While holding in the SET button, release the DISPLAY button. The minute readout will extinguish, leaving only the hours displayed. Release the SET button. Now, by depressing the DISPLAY button, the hours will run until the desired hour appears. Again depress the SET button. The hours will disappear and minutes will appear. Release the SET button. By depressing the DISPLAY button, the minutes will run until the desired minute appears. Once more depress the SET button until the minutes blank. Your watch is now set to the correct hours and minutes.

Date and Seconds: While holding in the DISPLAY button to show the date and seconds, simultaneously depress the SET button. Release the DISPLAY button. The seconds will soon disappear leaving only the date illuminated. Release the SET switch. Then by depressing the DISPLAY button, the date will run until the desired date appears. To set the seconds, depress the SET button. The date will disappear and the seconds will appear. Release the SET button. By depressing the DISPLAY button, the seconds will immediately go to :00 and will be in HOLD position. You must now again depress the DISPLAY button to restart the watch.

To Set Watch to the Exact Second: In order to set this watch to the exact second, the following additional procedure is advisable. Set the minutes two to three minutes ahead of the exact time. Then, when setting the seconds to :00, do not immediately reactivate the watch. Wait until the time standard catches up to the minutes that you previously set. At the exact second that the time standard reaches that minute, depress the DISPLAY button. This will start the timing circuit at the exact second desired.

FIVE FUNCTION: Hours, Minutes, Month, Date, Seconds **LED**

TYPE: 1

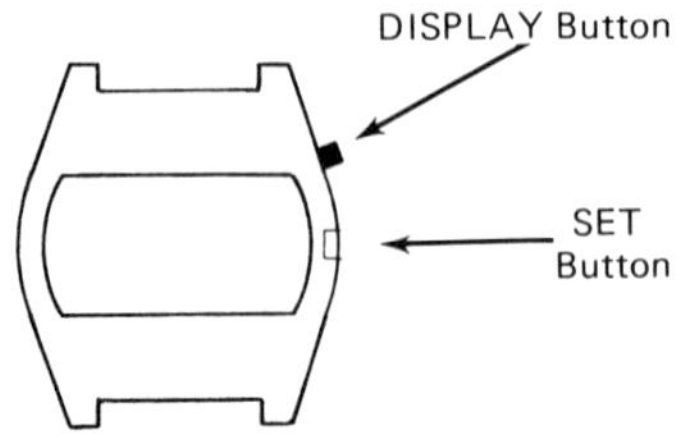

(A)

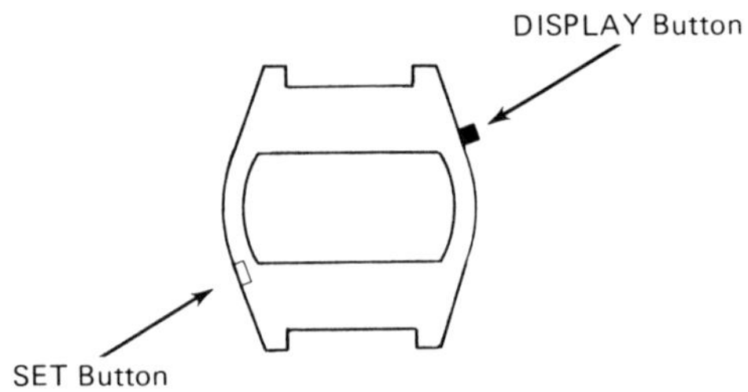

(B)

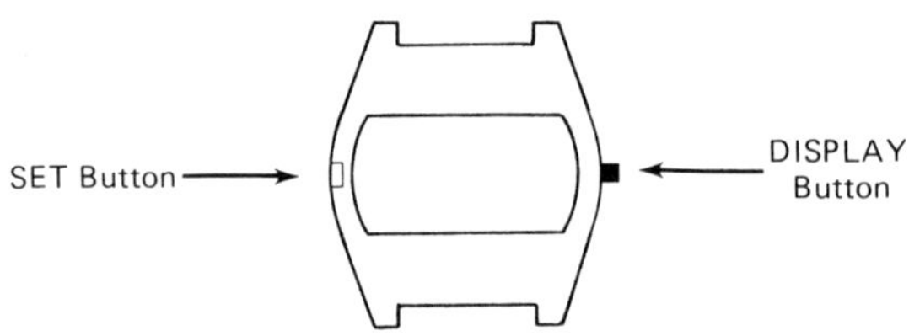

(C)

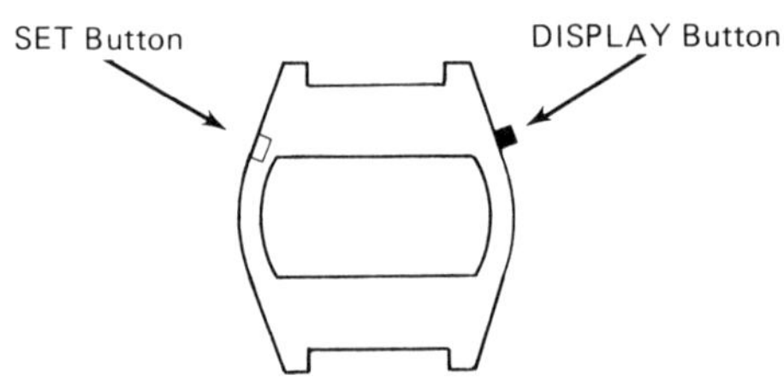

(D)

MANUFACTURERS:	RECOMMENDED BATTERY:	FIG.:	OTHERS:
Alcor	UC357	A	____________
American Microsystems, Inc.	—	—	____________
Bulova Computron			____________
Models 240, 229	Bulova 228	B or C	____________
Wyler	UC386, RW44	D	____________

OPERATION:

To show hours and minutes, depress the DISPLAY button. After the hours and minutes have been displayed for about 2 seconds the minutes and seconds will be displayed until the button is released. No colon is displayed. Depress the DISPLAY button two times in sequence, and after the hours and minutes are shown, the month and date will appear.

SETTING INSTRUCTIONS:

Hours or Time Zone Change: Depress and release the SET switch once to display hours and A (for A.M.) or P (for P.M.). Hours will be advanced by one whenever the DISPLAY button is depressed once. While the hours are advancing, the minutes and seconds counters continued to keep correct time. If you do not want to change any other setting (minutes, seconds, month, or date), just depress the SET switch five more times. This display will blank out in the normal mode, and the minutes and seconds will continue to run as before.

Minutes and Seconds: A second depression of the SET switch displays minutes and second. Depressing the DISPLAY button will cause the minute to advance by one. This step also resets the seconds counter to zero. To help you to obtain an accurate time setting, the minutes should be advanced to one minute behind the actual time.

Months: A third depression of the SET switch causes the months to be displayed. If you wish to change the month, depress the DISPLAY button the required number of times until the correct month appears.

Date: A fourth depression of the SET switch causes the months to be displayed. The date can be advanced, if desired, by pressing the DISPLAY button one by one when the correct date appears.

Display Test: Depress the SET switch a fifth time, and the display will show 88:88.

Normal: A sixth depression of the SET switch will return the display to normal.

FIVE FUNCTION: Hours, Minutes, Month, Date, Seconds **LED**

TYPE: 2

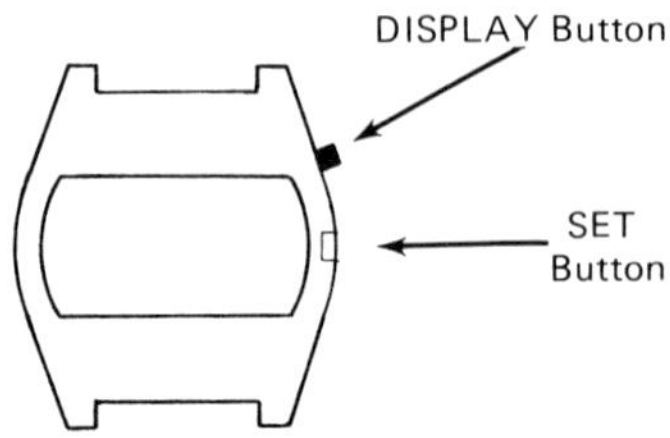

(A)

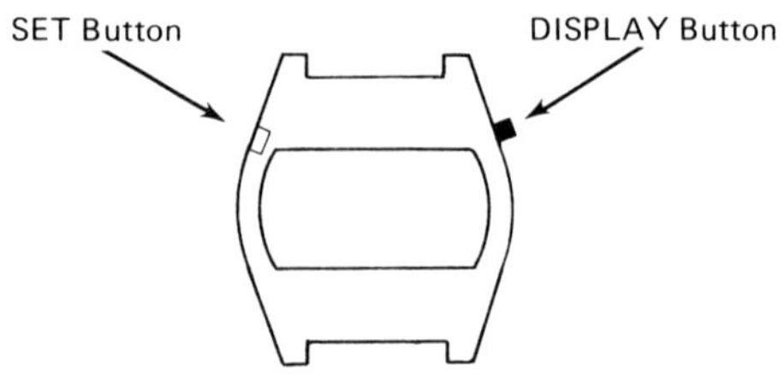

(B)

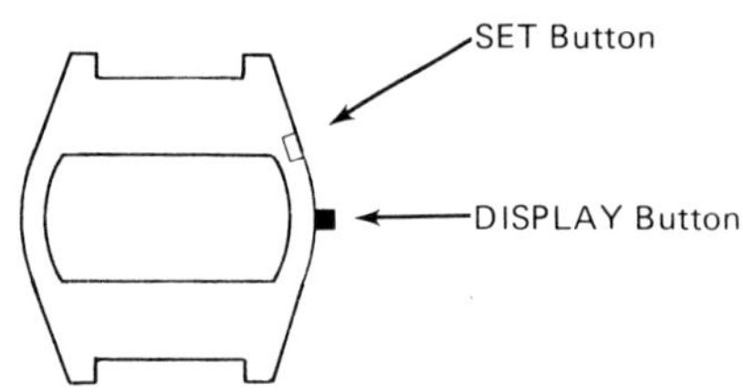

(C)

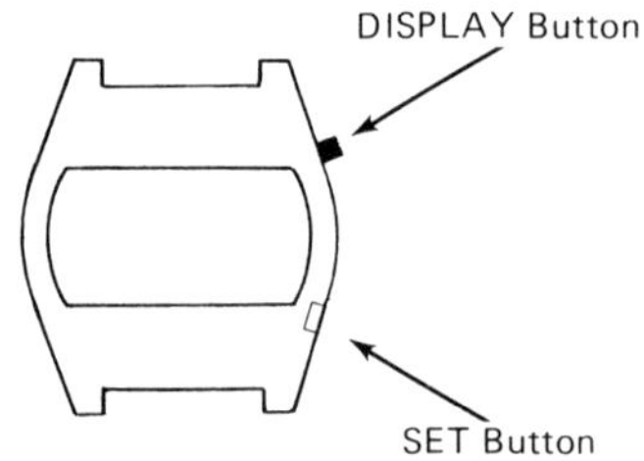

(D)

MANUFACTURERS:	RECOMMENDED BATTERY:	FIG.:
Alcor	UC357	A
American Microsystems	—	—
Armitron	*	B
Bulova Computron Model 253	RW44	B
EWT	UC386, RW44, M10L124	A
Girard Perrigaux	UC357	C
Marcel	UC357	D
Microsonic	(Mens-UC357, RW22, M10L14; Ladies-UC392, RW47)	A
Texas Inst.	RW44	D
Unitrex	UC392	A
Windert WF-5C	—	A
Windern WN-5C	—	D

*Engraved on inside of back

OPERATION:

To show hours and minutes, depress the DISPLAY button and release. Hours and minutes, separated by a colon, will be displayed. To show seconds, depress and hold the DISPLAY button. After the hours and minutes have been displayed, the seconds will be displayed until the button is released. No colon is displayed. For month/date, press the DISPLAY button twice quickly. After the hours and minutes are shown, the month and date will appear.

SETTING INSTRUCTIONS:

Hours or Time Zone Change: Depress and release the SET button once to display hours and A (for A.M.) or P (for P.M.). Depress the DISPLAY button to cause hours to advance. Release the button when the correct hour is reached. While the hours are advancing, the minutes and seconds counters continue to keep correct time. If you do not want to change any other setting (minutes, seconds, month or date), just depress the SET switch four more times. The display will blank out in the normal mode, and the minutes and seconds will continue to run as before.

Minutes: A second depression of the SET button displays minutes only. Depressing and holding the DISPLAY button will cause the minutes to advance continuously. This step also resets the seconds counter to zero. To allow you to obtain precise synchronization with an accurate time source, the minutes should be advanced to one minute past the actual time.

Months: A third depression of the SET button causes the months to be displayed. If you wish to change the month, push down and hold the DISPLAY button, releasing it when the correct month appears.

Date: A fourth depression of the set switch causes the date to be displayed. The date can be advanced, if desired, by pressing the DISPLAY button and releasing it when the correct date appears.

End of Setting Sequence: If the minutes had not been advanced, depressing the SET button one more time causes the display to go off, and this watch is now keeping time normally. Pressing the SET button will cause a blinking bar to appear if the minutes had been advanced, and the watch had been put into a zero (.00) seconds hold mode. As soon as the seconds of time standard come to zero, push the DISPLAY button. This will cause the hours and minutes to be displayed and your watch is now operating normally.

FIVE FUNCTION: Hours, Minutes, Month, Date, Seconds **LED**

TYPE: 3

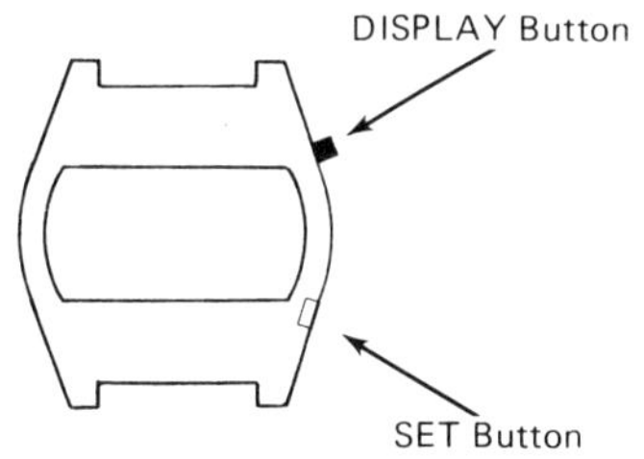

(A)

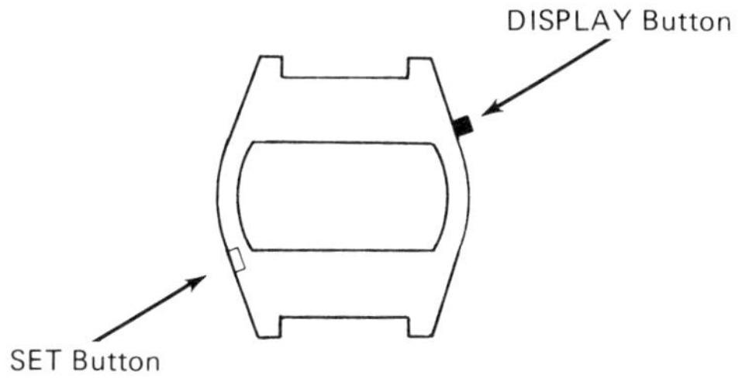

(B)

MANUFACTURERS:		RECOMMENDED BATTERY:	FIG.:
Advance	Majesty		
Arnex Quartz Ladies H5	Marcel	All Ladies:	A
Arnex Quartz Mens H1	Mercury Time	UC386, RW44	
Belair	Microsonic		
Benrus	Microsonic Ladies 5F	All Mens:	B
Concord	Timex*	UC386, RW44	
Croton	Westclox		
Duratime	Windert WH-5F-1		
Elgin	Windert WHL-5F-1		
Helbros	Zodiac		
Innovative Time			
Jupiter			

OPERATION:

To show hours and minutes, depress and release the DISPLAY button. To show seconds, depress and hold the DISPLAY button. First the hours and minutes will be displayed; then the seconds will appear. To turn the seconds display off, release button. To show month and date, simply press the DISPLAY button twice in rapid succession. First, the hours and minutes display will appear; then the month and date readout will come on. To keep the month and date display operational, hold the DISPLAY button in the depressed position. To turn the month and date display off, release this button.

SETTING INSTRUCTIONS:

Note: Must be followed in sequence, as outlined below.

Month: Depress the SET button to cause the month display and colon to appear and remain. Press the DISPLAY button to advance the display to the proper month. At the instant the proper month is reached, release this button.

Date: Depress the SET button a second time. This will cause the colon and date display to appear and remain. Press the DISPLAY button to advance the display to the proper date. At the instant the proper date is reached, release this button.

Hours: Depress the SET button a third time. This will cause the hours and colon display to appear and remain. Press the DISPLAY button to advance to the proper hour. (*NOTE:* A double colon, after the hour number, indicates A.M. time; a single colon after the hours number indicates P.M. time. Be sure to select the correct A.M. or P.M., as this will control the date change at 12 midnight.) At the instant the proper hour is reached, release this button.

Minutes: Depress the SET button a fourth time. This will cause the minutes display to appear and remain. Press the DISPLAY button to advance the display to the proper minute. At the instant the proper minute is reached, release this button.

Normal Timekeeping: Depress the SET button a fifth time. This will return your timepiece to its normal timekeeping mode and cause the display area to become blank. Press the DISPLAY button and you will return to hours and minutes timetelling.

Seconds: These are synchronized with the minutes display and automatically reset to zero-zero (00) when the minutes display was set as noted above.

**Note:* Timex differs only by showing minutes with seconds; not just seconds alone.

FIVE FUNCTION: Hours, Minutes, Month, Date, Seconds **LED**

TYPE: 4

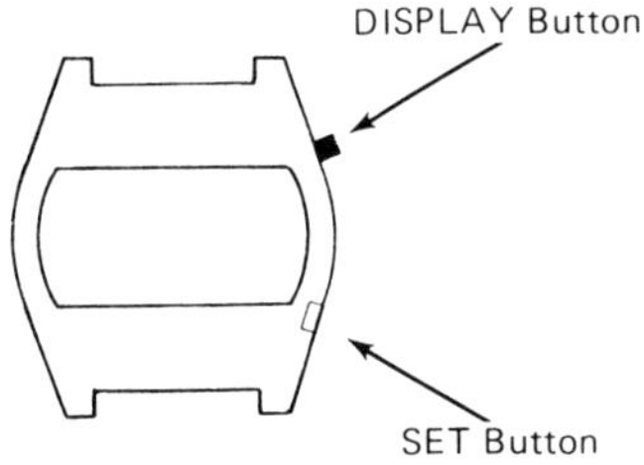

(A)

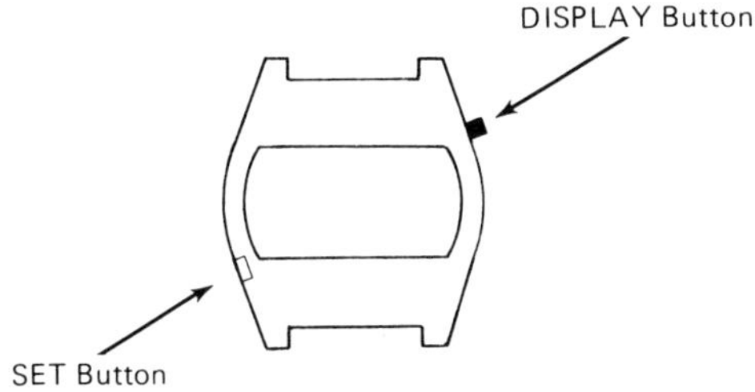

(B)

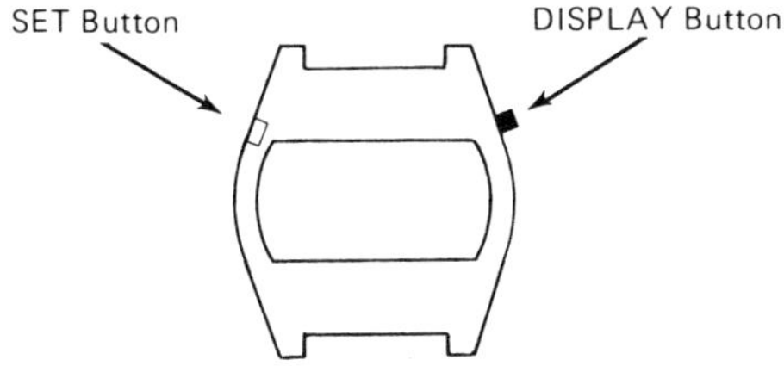

(C)

MANUFACTURERS:	RECOMMENDED BATTERY:	FIG.:
Arnex Quartz Mens E	—	A
Fairchild	UC357	A
Hughes	—	B
Quasar	*	C
Speidel	RW44, UC386	A
Timeband	(Mens—RW44, UC386, M10L124; Ladies—RW47, UC392, M10L125)	A
Waltham	RW44	A

OPERATION:

Hours and minutes can be displayed by one depression and release of the DISPLAY button. Month and date can be displayed by two successive depressions and release of the DISPLAY button, as long as the second depression occurs 1.5 seconds or less after release of the first depression. A third successive depression and release 1.5 seconds or less after release of the second depression will cause the seconds display to appear. All displays will be shown for 1.5 seconds after releasing the DISPLAY button.

SETTING INSTRUCTIONS:

Note: Must be followed in sequence outlined below.

Month: To set the month, depress and release the SET button once and the month will be displayed. Push the DISPLAY button, advance to the proper month and release button.

Date: Depress and release the SET button a second time and the date will be displayed. Push the DISPLAY button, advance to the correct date, and release button.

Hours: Depress and release the SET button a third time, and the hours, colon, and "A" or "P" will be displayed. Push the DISPLAY button, advance to the correct hour and A.M. or P.M., and release the button. (*Note:* "A" or "P" hours time must be set correctly to allow the date/month to change at 12:00 midnight.)

Minutes: Depress and release the SET button a fourth time, and the colon and minutes will be displayed. Advance the display to the correct minute by pushing the DISPLAY button. When correct minute appears, release button. If setting the watch to a time standard, advance minutes to the next minute the time standard will arrive at, and proceed to the next stop below.

Time Hold Mode: Depress and release the SET button a fifth time, at which time the display will be blank. The watch is now in the time hold mode—timekeeping will begin again as soon as the DISPLAY button is depressed. This allows for accurate setting to your time standard and minutes/seconds synchronization. (*Note:* the timepiece enters this hold mode only if minutes have been reset; if only month, date or hours have been previously reset, the watch will continue running normally, not enter the time hold mode, and the fifth successive depression and release will return the watch to normal operation.)

FIVE FUNCTION: Hours, Minutes, Seconds, Day, Date **LED**

TYPE: 5

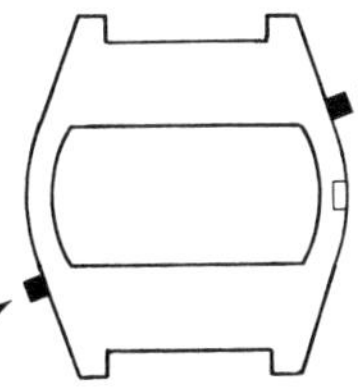

MANUFACTURERS:	RECOMMENDED BATTERY:	OTHERS:
Avatar S600	RW44, UC386, M10L124	__________
Litronix Series 6000	RW44, UC386, M10L124	__________
Windert WL-5A-2B	RW44, UC386, M10L124	__________

OPERATION:

Hours and minutes are displayed by pressing DISPLAY button #1. The hours and minute display will remain lit for approximately 1½ seconds after the button is released. To read seconds, continue to hold DISPLAY button #1 down until the hours and minutes changes to a seconds display. The seconds display will remain lit as long as this button is depressed. Day and date are displayed by depressing DISPLAY button #2. The day of the week will be displayed in alpha characters (SU for Sunday, MO for Monday, etc.). The day/date display will remain lit for approximately 1½ seconds after the button is released. (*Note:* Pressing DISPLAY button #1 and DISPLAY button #2 simultaneously will advance the date reading.)

SETTING INSTRUCTIONS:

Day: Depress DISPLAY buttons #1 and #2 and the SET button simultaneously until the correct day is displayed. Immediately release all buttons, and the day is set.

Date: Depress DISPLAY buttons #1 and #2 simultaneously until the correct date is displayed. Immediately release both buttons and the date is set.

Hours: Set hours by depressing DISPLAY button #1 and SET button simultaneously until the A.M./P.M. indicator is correct (one dot (.) between hours and minutes indicates A.M. and two dots (:) indicate P.M.) and the hours is set. Immediately release the buttons and the hour is set. Note that if the A.M./P.M. indicator is not correct, the day and date will advance at the wrong time of day.

Minutes: Set minutes by depressing DISPLAY button #2 and the SET button simultaneously until the correct minute is displayed, and allow the display to then roll to the next minute (in effect, the watch will now read one minute fast). Release DISPLAY button #2 immediately.

Seconds: To set seconds, press the SET button and hold it until the next full minute time signal occurs. Immediately release the button. This action automatically resets SECONDS to zero, and completes the setting of minutes and seconds.

FIVE FUNCTION: Hours, Minutes, Seconds, Day, Date **LED**

TYPE: 6

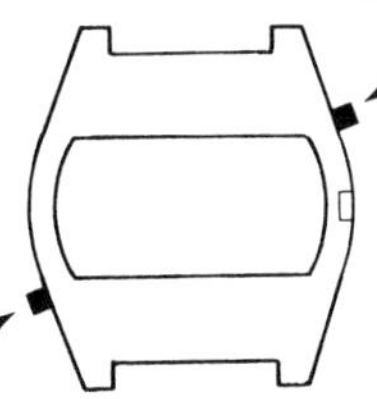

MANUFACTURERS:	RECOMMENDED BATTERY:	OTHERS:
Windert WL-5A	—	

OPERATION:

To show hours and minutes, depress DISPLAY button #1. To show day of the week and the date, depress DISPLAY button #2. To show seconds, depress DISPLAY buttons #1 and #2 at the same time.

SETTING INSTRUCTIONS:

Minutes: First, if split second accuracy is wanted, synchronize the seconds to an accurate time signal, by setting minutes to the next minute coming up. Press DISPLAY button #1 and the SET button to set the minutes. Release DISPLAY button #1, but continue holding the SET button. At the time signal, repress DISPLAY button #1 and release both buttons instantly. Your watch is now set for minutes and seconds.

Hours: Press only the SET button. A small dot in the display indicates A.M. hours. Two small dots indicate P.M. hours. Set the hours accordingly, otherwise, the day of the week will not change at midnight.

Day of the Week: Set day of week by pressing both DISPLAY buttons #1 and #2 and the SET button at once.

Date: Depress DISPLAY button #2 and the SET button.

FIVE FUNCTION: Hours, Minutes, Month, Date, Seconds **LED**

TYPE: 7

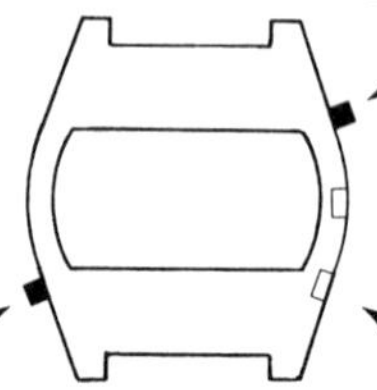

MANUFACTURERS:	RECOMMENDED BATTERY:	OTHERS:
Citizen CQ	UC386	

OPERATION:

Press DISPLAY button #1 to get hours and minutes, and hold to display seconds. Press DISPLAY button #2 to get month and date. A hyphen between month and date indicates A.M.; a comma between month and date indicates P.M.

SETTING INSTRUCTIONS:

Minutes: Press SET button #1 to advance minutes and to zero seconds. The watch can be synchronized with a time tone by depressing DISPLAY button #1 to start seconds.

Hours: Press SET button #2 to advance hours one hour per push.

Date and A.M./P.M.: Press DISPLAY buttons #1 and #2 and SET button #2 to advance date and A.M./P.M.

Month: Press DISPLAY button #2 and SET button #2 to advance month.

FIVE FUNCTION: Hours, Minutes, Month, Date, Seconds **LED**

TYPE: 8

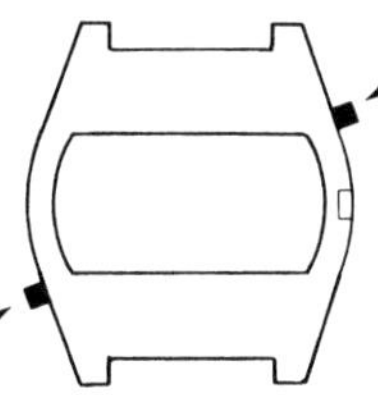

MANUFACTURERS:

Advance
Arnex Quartz Ladies
Arnex Quartz Mens
Belair
Benrus
Concord
Croton
Duratime
Elgin
Helbros
Hughes 2 Button
Innovative Time
Jupiter
Majesty
Marcel
Mercury Time
Microsonic
Microsonic Ladies
Westclox
Wittnauer Polara
Windert WH-5F-2
Windert WHL-5F-2
Zodiac

OTHERS:

Recommended Battery: On all watches listed—UC357, RW22.

OPERATION:

To display the time in hours and minutes, depress DISPLAY button #1. The hours and minutes will remain visible for a short period of time. If DISPLAY button #1 is held down continuously, hours and minutes will disappear and a flashing seconds will begin, and remain in view so long as the button is depressed. To display month and date, depress DISPLAY button #2. Note that the month and date are separated by a dash (—), while hours and minutes are separated by a colon (:).

SETTING INSTRUCTIONS:

Month: Depress the SET button once and the month number will start flashing. The watch is now in the month-set mode. To change the month, depress DISPLAY button

#1, which advances the months, and release when the desired month number appears. The correct month is now set into the watch.

Date: Depress the SET button a second time and the date will appear. This indicates that the watch is in the date-set mode. Depress DISPLAY button #1 and hold until the correct number for the date appears. This watch is programmed to adjust itself automatically for 29, 30, and 31 day months.

Hours: Depress the SET button a third time. Hours will begin to flash, indicating the hours-set mode. To change the hours, depress DISPLAY button #1 until the correct hour is displayed. If the watch is set to A.M. hours, the colon will include a dash. During the P.M. hours, the dash disappears and you see just the colon.

Minutes: Select the minutes-set mode by depressing the SET button until minutes are flashing. Using an accurate time source, listen for the next full minute time signal. At this instant, depress DISPLAY button #1 and hold it down until the correct minute flashes into view. Release the DISPLAY button. The watch is now set to the nearest second. As a special feature, this watch delays the start of advancing minutes by two or three seconds. During this interval, the DISPLAY button #1 can be pushed and released, returning seconds to zero but not advancing minutes.

FIVE FUNCTION: Hours, Minutes, Seconds, Month, Date **LED**

TYPE: 9

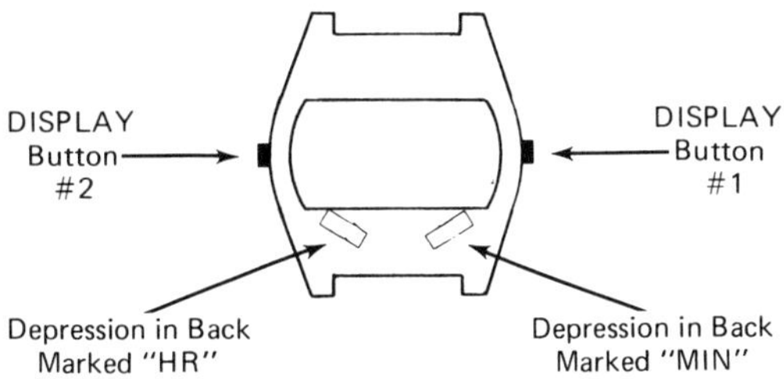

MANUFACTURERS:	RECOMMENDED BATTERY:	OTHERS:
Pulsar	UC355, RW25	
Omega	UC355, RW25	

OPERATION:

Press DISPLAY button #1 to display hours and minutes. Continue to depress and seconds will appear in 1¼ seconds and will remain as long as the button is depressed. Press DISPLAY button #2 to display month and date.

SETTING INSTRUCTIONS:

Note: A magnetic timeset bar is stored in the clasp of the bracelet or strap for setting purposes.
Hours: To change the hour setting, take the timeset bar from the bracelet clasp and insert it into the recess on the case back marked "HR." This causes the displays to light up and the hours to advance. Remove the timeset bar when the proper hour has been reached.

Minutes: To reset minutes and seconds, insert the timeset bar in the recess on the case back marked "MIN." This causes the displays to light up and minute digits to advance. It also automatically resets seconds to 00. Set the time to the upcoming minute and quickly remove the timeset bar. At an accurate time tone, depress DISPLAY button #1 to start timekeeping.

Date and A.M./P.M. Indicators: Simultaneously press both DISPLAY buttons (#1 and #2), then place timeset bar in the "HR" recess in the case back. This will cause the displays to light up and advance the date. At the same time, as each date number changes, the A.M. and P.M. indicators (upper and lower dots between month and date numbers) will alternately light up. Remove the timeset bar when the proper date appears and when the A.M. or P.M. indicator is correct.

Month: Press DISPLAY button #2, then insert timeset bar in the "HR" recess in the case back. Again, the displays will light up and the month will advance. Remove the timeset bar when the correct month appears.

After setting time and date, replace the timeset bar in the bracelet or strap clasp.

FIVE FUNCTION: Hours, Minutes, Seconds, Day, Date **LED**

TYPE: 10

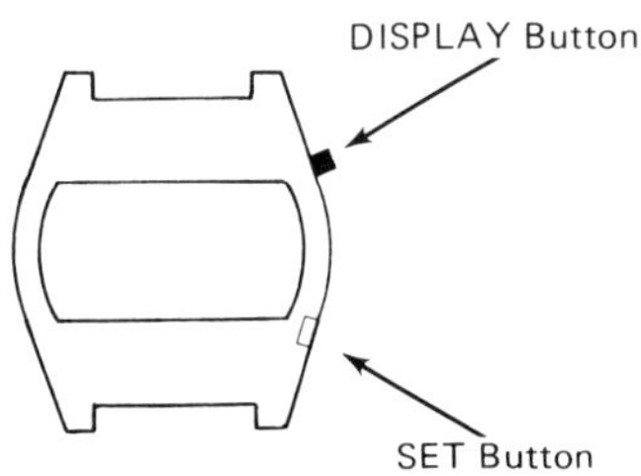

MANUFACTURERS:	**RECOMMENDED BATTERY:**	**OTHERS:**
Cronus		

OPERATION:

The hours and minutes are displayed by depressing the DISPLAY button. Holding the DISPLAY button in will cause the timepiece to roll to seconds. Two rapid successive depressions of the DISPLAY button will display day of week and date.

SETTING INSTRUCTIONS:

Date: When the SET button is pressed once, the two right digits will light up to show the date. By pressing the DISPLAY button, the correct date of the month can be set.

Hour: When the SET button is pressed a second time, the two left digits will light up to show the hour. By pressing the DISPLAY button repeatedly, hours can be advanced to the right time. The colon unlit tells you the time shown is A.M. The colon lit tells you the time shown is P.M.

Day: When the SET button is pressed a third time, the two left digits will light to show the day of the week. By pressing the DISPLAY button repeatedly, the correct day of the week can be set.

Minutes: When the SET button is pressed a fourth time, the two right digits will light up to show minutes. By pressing the DISPLAY button repeatedly, minutes can be advanced to the right time.

Seconds: When the SET button is pressed a fifth time, the two right digits will light up to show seconds. You will note that the digits count continually. When the *actual* time is an even minute, press the DISPLAY button, which resets the digits to 00 and begins the seconds to count.

FIVE FUNCTION: Hours, Minutes, Seconds, Day, Date **LED**

TYPE: 11

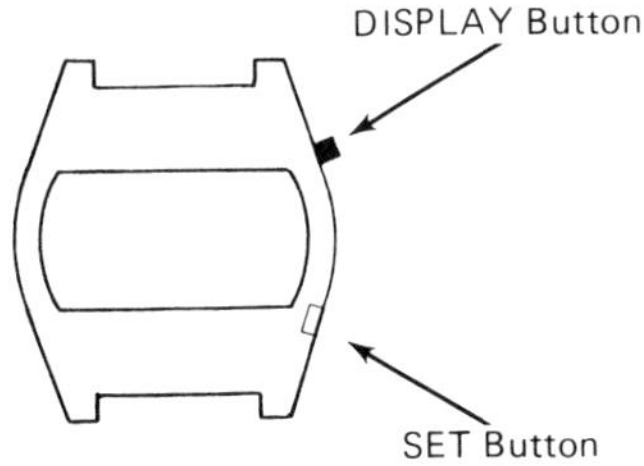

MANUFACTURERS:	**RECOMMENDED BATTERY:**	**OTHERS:**
Litronix Ladies	UC392, RW47	

OPERATION:

The hours and minutes are displayed by depressing the DISPLAY button. Holding the DISPLAY button in will cause the timepiece to roll to seconds. Two rapid successive depressions of the DISPLAY button will display day of week and date.

SETTING INSTRUCTIONS:

Day: Depress the SET button and hold until display rolls to day of week. This places the timepiece in a holding mode and will remain in hold until setting is complete. When day of week displays, release the SET button and press the DISPLAY button. Hold until correct day appears. Immediately release button and day of week is set.

Date: Depress the SET button a third time and the hour and the day of month will display. Release the SET button immediately and press the DISPLAY button. Hold until the correct date displays, then release.

Hour: Depress the SET button a third time and the hour and an "A" or "P" will display. Release the SET button, press the DISPLAY button and hold until the correct hour displays, then release the button. The "A" indicates A.M., "P" indicates P.M.

Minutes: Depress the SET button a fourth time and minutes will display. Release the set button immediately, press the DISPLAY button and hold until the display reads one minute ahead of real time, then release the button.

Seconds: If minutes were reset, wait until the next full minute signal, depress the SET button a fifth time, and timekeeping will restart. The timepiece is now in the normal operating mode.

FIVE FUNCTION: Hours, Minutes, Seconds, Day, Date **LED**

TYPE: 12

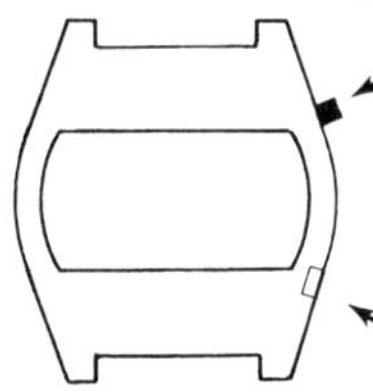

MANUFACTURERS:	RECOMMENDED BATTERY:	OTHERS:
Novus	RW22, UC357, M10L14 (*Note:* On thin battery models —RW44, UC386, M10L124)	

OPERATION:

The hours and minutes are displayed by depressing the DISPLAY button. Holding the DISPLAY button in will cause the timepiece to roll to seconds. Two rapid successive depressions of the DISPLAY button will display day of week and date.

SETTING INSTRUCTIONS:

Hours: Depress the SET button once and release. Hours on left of colon and "A" or "P" (indicates A.M. or P.M.) on right of colon display will stay on. Depress the DISPLAY button. Hours will advance at the rate of 2 per second. When correct hours and "A" or "P" appears, release the DISPLAY button. The hour display will go off 5 seconds after release and the watch returns to the normal run state if no other button is pushed.

Minutes: Depress the SET button twice and release. Minutes display will stay on. Depress the DISPLAY button, seconds will set to 00, minutes will advance at the rate of 2 per second. When correct minutes appear, release the DISPLAY button. After 5 seconds the display automatically goes into a Hold state with hours and minutes flashing and seconds held to 00. When the time signal indicates exactly the displayed time, depress the DISPLAY button and release it, which starts the counter counting seconds, minutes, and hours at normal rate. The watch is now set to the precise second.

Day of Week: Depress the SET button three times and release, day of the week will be displayed. Depress the DISPLAY button, days of the week will advance from the display day. Release the DISPLAY button when correct day appears. The day display will go off 5 seconds after button release and the watch returns to the normal run condition if no other button is pushed.

Date: Depress the SET button four times and release. Day of month will display. Depress the DISPLAY button and days will advance, release the DISPLAY button when correct day appears. The display will go off 5 seconds after button release and the watch returns to the normal run condition if no other button is pushed.

Month: Depress the SET button five times and release, "M" (for month) and number of month will be displayed. Depress the DISPLAY button, number of months will advance. ("M" will remain.) Release the DISPLAY button when correct month appears. The display will go off 5 seconds after button release and goes into normal run condition if no other button is pushed. The month counter provides you with a 4 year calendar but is only displayed during SET MONTH. The date will count 28 days in February, 30 in April, June, September, and November, and 31 in the remaining months.

FIVE FUNCTION: Hours, Minutes, Month, Date, Seconds **LED**

TYPE: 13

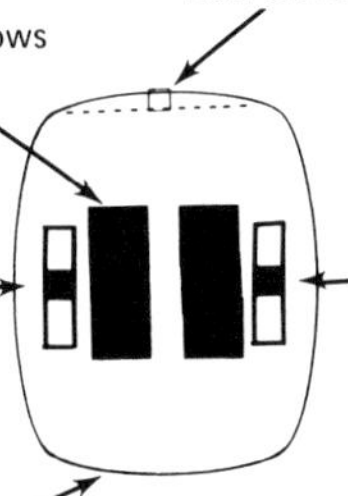

MANUFACTURERS:

Ragen Synchronar

OTHERS:

____

____

____

Recommended Battery: Must be returned to manufacturer for battery replacement

OPERATION:

Holding the timepiece with the front edge display facing you, slide the right hand slide bar away to display time in hours and minutes. Slide this bar toward you to display date in months and days (for example, July 25th would display 7:25). Slide the left hand slide bar toward you to display the passing seconds.

SETTING INSTRUCTIONS:

Hours: Holding the timepiece with the front edge display facing you, slide the left hand slide bar away from you to advance the watch one hour for changes in time zone for the daylight savings adjustment. Advancing hours are not displayed when the time slide bar is actuated (this slide bar has a stronger spring return than the others to prevent accidental actuation). Advancing this slide bar will not affect the calendar reading.

Minutes and Date: Holding the timepiece with the rear edge facing you, slide the rear slide bar right, to advance the day in the calendar section to the desired day number. This adjustment will not affect the month and year reading, as it will merely repeat the days within the month displayed. Slide the bar to the left to advance the time one minute and by holding it keep the seconds at zero while awaiting an exact time signal thus synchronizing the watch very closely by releasing this slide bar at the time of the signal.

SOLAR POWER SYSTEM:

Light, falling upon the upper surface of the solar cells is converted into electricity by its solar cells. This energy is stored in long life nickel cadmium cells, so that normal operation continues in the dark, up to six months. Beyond a predetermined point of use, the digital display is automatically kept off to conserve the remaining energy for timekeeping until sufficient exposure to light revives it. A fully charged battery should function for six months without exposure to light when used normally, ie. (25 display actuations per day of ½ seconds each.) If the average exposure to light accumulates more power than average use consumes, no special exposure provisions need be observed by the owner. Usually, approximately 15 minutes of daylight daily should suffice to maintain peak power. The display brightness will automatically adjust to ambient light or storage cell conditions. Exposure to sunlight will completely charge exhausted storage cells in 14 hours or less, whereas, an incandescent table lamp may take several times longer to do the same.

Caution: Do not expose the watch to overheating when leaving it to charge in the sun. Place it on a cool surface.

SIX FUNCTION: Hours, Minutes, Seconds, Month, Date, Day **LED**

TYPE: 1

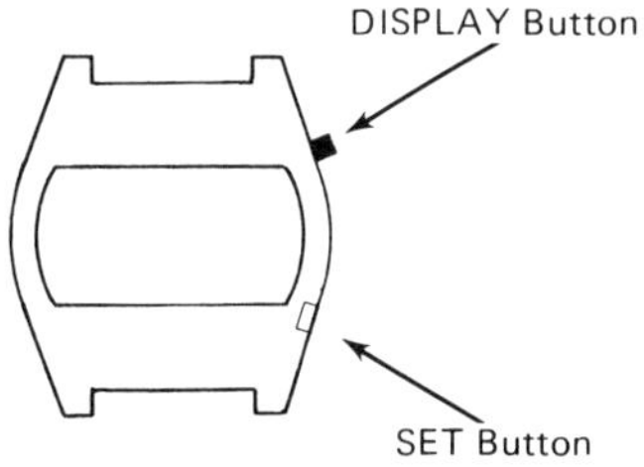

(A)

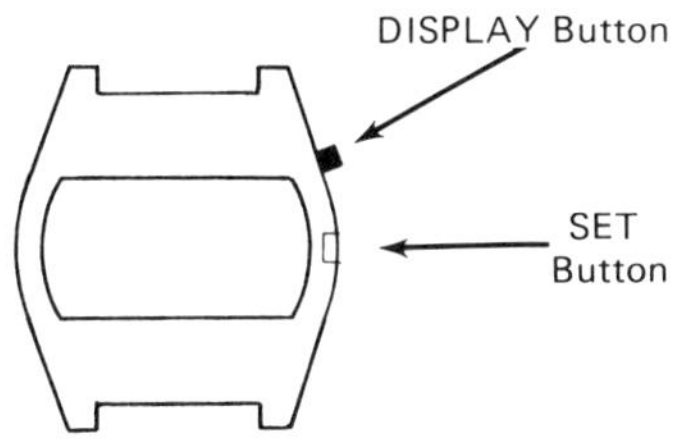

(B)

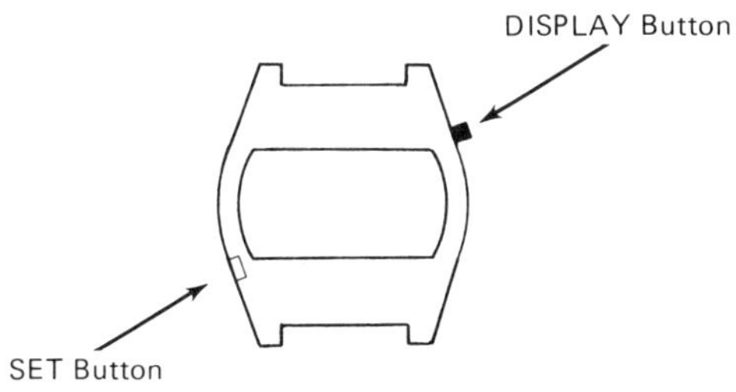

(C)

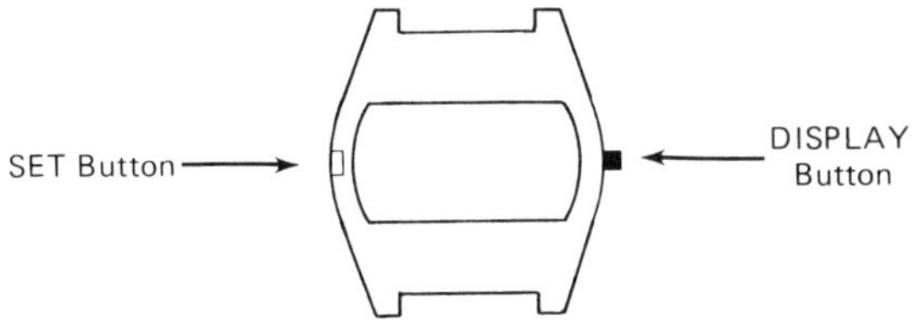

(D)

MANUFACTURERS:	RECOMMENDED BATTERY:	FIGURE:
Advance	UC357	A
Alcor	Mens–M10L14, UC357; Ladies–M10L125, UC392	B
Bulova Computron 2404, 2244	Bulova 228	C or D
Mercury Time	UC357, M10L14	A
Microsonic Single Button	Mens–UC357, RW22, M10L14; Ladies–UC392, RW14	B
Windert WF-6A	—	B
Wyler	RW44	A

OPERATION:

For hours and minutes, depress the DISPLAY button and release. The hours and minutes will be displayed for about 2 seconds. To read seconds, depress the DISPLAY button and hold. Hours and minutes will be displayed for about 2 seconds, then the display will automatically change for seconds and remain displayed until the button is released. Month and date are shown by depressing the DISPLAY button once and release. This displays hours and minutes. While the display is still on, again depress the DISPLAY button. This will display the month and date, which will remain for about 2 seconds. To show day of week, depress the DISPLAY button once, release and press again to obtain month and date; retain pressure on the DISPLAY button, and after approximately 2 seconds, the day of the week will appear.

SETTING INSTRUCTIONS:

Hours: Depress the recessed SET button once and release. This will display the Hours and an "A" (for A.M.) or a "P" (for P.M.). Depress the DISPLAY button and the hours will advance twice each second after a 1.5 second delay. Release at the proper count and A.M. or P.M. position.

Minutes: Depress the SET button a second time. This will display minutes. Depress the DISPLAY button and the minutes will advance twice each second after a 1.5 second delay. Release at the proper count.

Seconds: Seconds are set while in the minutes set mode. Depressing the DISPLAY button resets the seconds counter to zero and initiates the counting of seconds. Thus, depress sharply the DISPLAY button at the "sound of the tone" and seconds will be set exactly.

Months: Depress the SET button a third time. This will display the month. Depress the DISPLAY button and the months will advance twice each second after a 1.5 second delay. Release at the proper count.

Date: Depress the SET button a fourth time. This will display the day of the month. Depress the DISPLAY button and the date will advance twice each second after a 1.5

second delay. Release on the proper count. The date will automatically count to 28, 30 or 31, depending on the month; the module has been programmed for 4 years. (February is always 28.)

Day of Week: Depress the SET button a fifth time. This will display the day of week. Depress the DISPLAY button and the day will advance twice each second after a 1.5 second delay. Release on the proper day.

Important: The watch will remain in any set mode with digits on until advanced to another set mode or returned to normal operation. Therefore, the setting of any function requires a complete cycle of six depressions of the SET button to return the watch to normal operation.

SIX FUNCTION: Hours, Minutes, Seconds, Day/Date, Month/Date **LED**

TYPE: 2

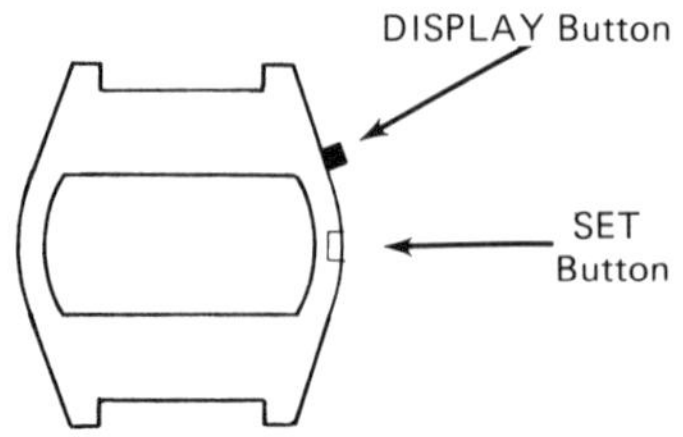

(A)

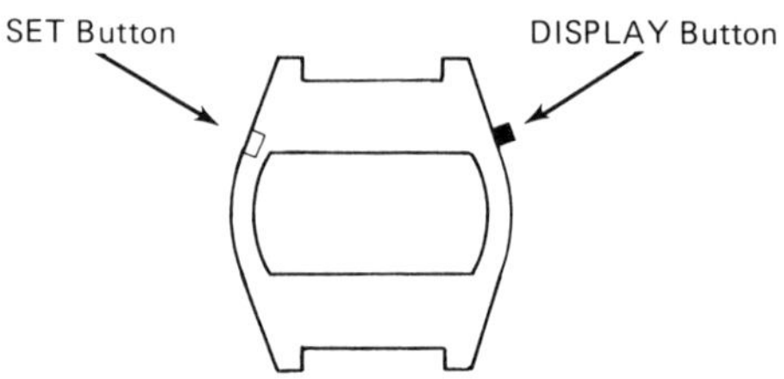

(B)

MANUFACTURERS:	**RECOMMENDED BATTERY:**	**FIG.:**	**OTHERS:**
Bulova #260	UC386	A	________
Quasar	*	B	________
			________

*Engraved on inside of case back.

OPERATION:

To read the correct time—hours, minutes and seconds, depress the DISPLAY button one time. Please note that a dot separates the hour and minutes. Seconds appear after hours and minutes. To read day/date and month/date, depress DISPLAY button twice and hold. Please note that there is no dot separating the day from the date. After day and date, month and date will appear.

SETTING INSTRUCTIONS:

Month: Depress the SET button and release. The digits for the month (1 through 12) will appear. If the month is correct, proceed with the next step. If the incorrect numeral for the month appears, depress the DISPLAY button and retain until the correct figure appears. Immediately release the DISPLAY button.

Day: Depress SET button and release. The letters for the day of the week (M, TU, W, TH, F, SA, SU) will appear. If the incorrect letters appear, depress the DISPLAY button and retain until the correct day appears. Immediately release the DISPLAY button.

Date: Depress the SET button again, and the numerals for the date will appear. If this is incorrect, depress the DISPLAY button and retain until the correct date appears.

Hours: Depress the SET button again and the hour digits will appear together with an "A" (for A.M.) or "P" (for P.M.). If the hours, A.M. or P.M. designation is incorrect, press and retain the DISPLAY button until they appear correctly. Continue with the next step.

Note: It is important that the correct A.M. or P.M. designation is used, otherwise the date will change at noon instead of midnight.

Minutes: In the setting of minutes, the seconds are automatically reset to zero, therefore it will be necessary to advance the minutes one count ahead of actual time. Depress the SET button again and the minutes will appear. In order to set your watch to the exact minute and seconds, you must advance the minute numerals one minute ahead of the actual time by depressing the DISPLAY button. When the minutes have been set, release the DISPLAY button. At time tone, depress SET button a sixth time. Display will blank, seconds will start counting and watch is back in normal timekeeping mode.

SIX FUNCTION: Hours, Minutes, Seconds, Month, Date, Day **LED**

TYPE: 3

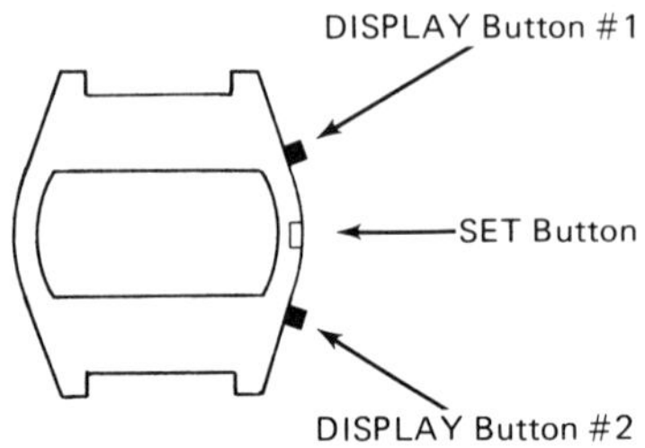

MANUFACTURERS:	**RECOMMENDED BATTERY:**	**OTHERS:**
Datatime	UC386 or equivalent	______________
		______________
		______________

OPERATION:

To display hours and minutes, depress DISPLAY button #1 and release. For seconds, depress DISPLAY button #1; after hours and minutes have been displayed, the seconds will be displayed until the button is released. To display month and date, depress DISPLAY button #2 and release. For day of the week, press and hold DISPLAY button #2; after the month and date have been displayed, the day of the week will be displayed until the button is released.

SETTING INSTRUCTIONS:

Month: Press the SET button once to enter month mode. The month will be displayed in the left hand side of digits. To advance at the rate of one month per second, depress DISPLAY button #2 until the desired month is reached.

Date: To set the date, press the SET button again to enter date mode. Press DISPLAY button #2 to advance the date. The date will always be displayed when in the date setting mode.

Day: To set day, depress the SET button again to enter the day mode. The day will appear in the right hand set of digits and will advance at the rate of one day per second when DISPLAY button #2 is pressed.

Hour: Press the SET button again to enter the hour mode. The hour will appear in the left hand set of digits with an "A" for A.M. or a "P" for P.M. in the right hand digits. It is necessary to set the hours correctly for A.M. or P.M. to maintain proper synchronization of the calendar functions. Press DISPLAY button #2 to advance hours at the rate of one hour per second.

Minutes: Again depress the SET button to enter the minutes mode. The minutes will appear in the right hand set of digits; however, until DISPLAY button #1 is pressed, timekeeping will continue. When DISPLAY button #1 is depressed, even momentarily, the seconds reset to zero and timekeeping is stopped until the SET button is depressed again. This puts the watch into a hold condition for synchronization with a time reference. To advance minutes at the rate of one minute per second, depress DISPLAY button #1 until the desired time is reached.

Note: In all setting positions, the selected mode will display the actual setting. That setting can be advanced, stopped and advanced again without affecting any of the other functions except for the minutes section as described above.

To return to normal timekeeping and display, press the SET button again when the time reference reaches zero seconds. The watch will be synchronized with the time reference.

SIX FUNCTION: Hours, Minutes, Seconds, Month/Date, Day/Date **LED**

TYPE: 4

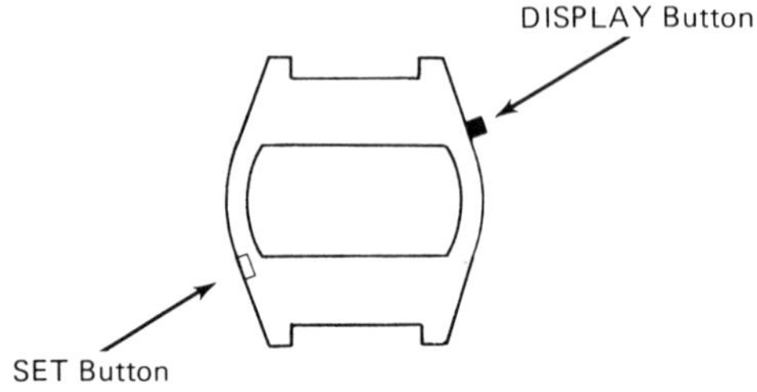

MANUFACTURERS:		**OTHERS:**
Advance	Innovative Time	
Arnex Quartz Ladies	Jupiter	
Arnex Quartz Mens	Longines Wittnauer	
Belair	Majesty	
Benrus	Marcel	
Concord	Mercury Time	
Croton	Microsonic	
Duratime	Timex	
Elgin	Westclox	
Helbros	Windert WH-6FA-1B	
Hughes 1 Button	Zodiac	

Recommended Battery: For all above watches—UC386, RW44

OPERATION:

To display the time in hours and minutes, depress the DISPLAY button. The hours and minutes will remain visible for a short period of time. If the DISPLAY button is held down continuously, hours and minutes will disappear and a flashing seconds display will begin. To display the day and date, press the DISPLAY button twice. If the button is held down continuously, the day and date will disappear and the month/date display will commence. For "hands-off" seconds, press the DISPLAY button 3 times. Seconds will continue to flash until the button is pressed a fourth time. If this is not done, the display automatically turns off after a minute or two to preserve battery life.

SETTING INSTRUCTIONS:

Year: Press the SET button once and the year number will start flashing. The watch is now in the "year mode." To change the year, hold the DISPLAY button down and the year number will begin to advance. When the desired number flashes into view, release the DISPLAY button. The correct year is now set into your watch. The year number is coded as follows:

FLASHING DISPLAY	REPRESENTS
0Y	Leap Year, e.g.; 1976, 1980, 1984, etc.
1Y	Year after Leap Year, e.g.; 1977, 1981, 1985, etc.
2Y	Two years after Leap Year, e.g.; 1978, 1982, 1986, etc.
3Y	Three years after Leap Year, e.g.; 1979, 1983, 1987, etc.

Once the correct year is set into the watch, it is never necessary to reset it to adjust for leap year variations.

Month: Press the SET button a second time and the month will begin to flash, indicating that the watch is in its "month set mode." To change this month, depress the DISPLAY button and hold it down until the correct number appears.

Date: Press the SET button a third time. The date will begin to flash, indicating the "date set mode." To change the date, depress the DISPLAY button until the desired date appears. The watch is programmed to adjust itself automatically for 28, 29, 30 and 31 day months. Consequently, resetting is never necessary to compensate for calendar variations.

Day: Press the SET button a fourth time and the day will start to flash. Select the correct day by pressing the DISPLAY button as before.

A.M./P.M.: Press the SET button a fifth time and the A.M. or P.M. designation will begin to flash. Choose either A.M. or P.M. by pressing the DISPLAY button. If the A.M./P.M. setting is incorrect, the date will change at noon instead of midnight.

Hours: Press the SET button a sixth time and the hour will begin to flash indicating the "hour set mode." Set the correct hour by depressing the DISPLAY button until the desired number appears.

Minutes: Press the SET button a seventh time and minutes will begin to flash. At the full minute signal, depreess the DISPLAY button and hold it down until the correct minute flashes into view. Your watch is now set to the nearest second.

SIX FUNCTION: Hours, Minutes, Seconds, Day/Date, Month/Date **LED**

TYPE: 5

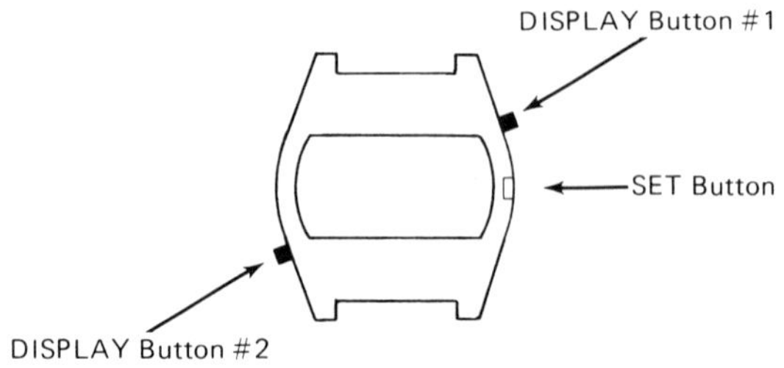

MANUFACTURERS:

Advance
Arnex Quartz Ladies
Arnex Quartz Mens
Belair
Benrus
Concord
Croton
Duratime
Elgin
Helbros
Hughes 2 Button
Innovative Time
Jupiter
Longines Wittnauer
Majesty
Marcel
Mercury Time
Microsonic
Timex
Westclox
Windert WH6–FA–2B
Zodiac

OTHERS:

____
____
____
____
____
____
____
____
____
____

Recommended Battery: On all above watches—UC357, RW22

OPERATION:

To display the time in hours and minutes, simply touch DISPLAY button #1. The hours and minutes will remain visible for a short period of time. If DISPLAY button #1 is held down continuously, hours and minutes will disappear and a flashing seconds display will begin. To display the day and date, simply depress DISPLAY button #2. If DISPLAY button #2 is held down continuously, the day and date will disappear and the month/date display will commence. For "hands-off" seconds, hold both buttons down for a moment and release. Seconds will continue to flash until either button is pressed. If this is not done, the display automatically turns off after a minute or two to preserve battery life.

SETTING INSTRUCTIONS:

The SET button is used to select specific watch functions for setting. The DISPLAY button #1 is used for the actual setting operation. Pressing the SET once starts the year number flashing, signifying that the year can be set. Pressing it again enables you to set the month; the third time, the date; the fourth, the day; etc.

Year: Press the SET button once and the year number will start flashing. The watch is now in the "year set mode." To change the year, simply hold the DISPLAY button #1 down and the year number will begin to advance. When the desired number flashes into view, release the DISPLAY button #1. The correct year is now set into the watch. The year number is coded as follows:

FLASHING DISPLAY	INDICATES
0Y	Leap Year, e.g.; 1976, 1980, 1984, etc.
1Y	Year after Leap Year, e.g.; 1977, 1981, 1985, etc.
2Y	Two Years after Leap Year, e.g.; 1978, 1982, 1986, etc.
3Y	Three years after Leap Year, e.g.; 1979, 1983, 1987, etc.

Once the correct year is set into the watch, it is never necessary to reset it to adjust for leap year variations.

Month: Press the SET button a second time and the month will begin to flash, indicating that the watch is in its month set mode. To change the month, press the DISPLAY button and hold it down until the correct number appears.

Date: Press the SET button a third time. The date will begin to flash, indicating the date set mode. To change the date, press the DISPLAY button #1 until the desired date appears. Your watch is programmed to adjust itself automatically for 28, 29, 30 and 31 day months. Consequently, resetting is never necessary to compensate for calendar variations.

Day: Press the SET button a fourth time and the day will begin to flash. Select the correct day by depressing DISPLAY button #1 as before.

A.M./P.M.: Press the SET button a fifth time and the A.M./P.M. designation will begin to flash. Choose either A.M. or P.M. by pressing the DISPLAY button #1. If the A.M./P.M. setting is incorrect, the date will change at noon instead of midnight.

Hours: Press the SET button a sixth time and the hour will begin to flash, indicating the hour set mode. Set in the correct hour by pressing the DISPLAY button #1, as before, until the desired number appears.

Minutes: Press the SET button a seventh time and minutes will begin to flash. At the time tone, press DISPLAY button #1 and hold it down until the correct minute flashes into view. The watch is now set to the nearest second.

SIX FUNCTION: Hours, Minutes, Seconds, Month/Date, Day **LED**

TYPE: Alarm

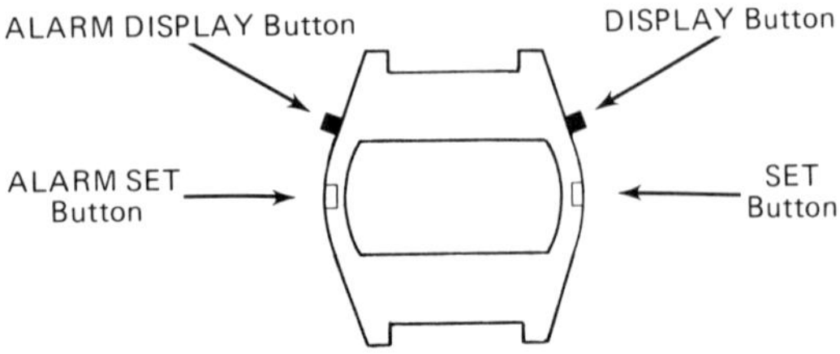

MANUFACTURERS:	RECOMMENDED BATTERY:	OTHERS:
Helbros	UC386, RW24 or 44	______
Windert	UC386, RW24 or 44	______
Unitrex	UC386, RW24 or 44	______

WATCH OPERATION:

Depress DISPLAY button and release for hours and minutes to be displayed. Hours and minutes will be displayed for about 2.0 seconds. To read seconds, depress the DISPLAY button and hold. After hours and minutes are displayed, the display will automatically change to seconds and remain displayed until button is released. Month and date are shown by depressing the DISPLAY button once and releasing. This displays hours and minutes. While the display is still on, again depress the DISPLAY button. This will display month and date, which will remain for about 2.0 seconds. To show day-of-week, depress the DISPLAY button once, release and press again to obtain month and date; retain pressure on DISPLAY button, and after approximately 2.0 seconds, the day-of-week will appear.

WATCH SETTING INSTRUCTIONS:

Hours: Depress the recessed SET button once and release. This will display the hours and an "A" (for A.M.) or a "P" (for P.M.). Depress the DISPLAY button and the hours will advance twice each second after a 1.5 second delay. Release at the proper count and A.M. or P.M. position.

Minutes: Depress the SET button a second time. This will display minutes. Depress DISPLAY button and the minutes will advance twice each second after a 1.5 second delay. Release at the proper count.

Seconds: To set the watch to the exact second is done while in the minutes set mode. Depressing the DISPLAY button resets the seconds counter to zero and initiates the counting of seconds. Therefore, depress sharply the DISPLAY button at the "sound of the tone" and seconds will be set exactly.

Month: Depress the SET button a third time. This will display the month. Depress the DISPLAY button and the months will advance twice each second after a 1.5 second delay. Release at the proper count.

Date. Depress the SET button a fourth time. This will display the day of the month. Depress the DISPLAY button and the date will advance at the rate of twice each second after a 1.5 second delay. Release on the proper count. The date will automatically count to 28, 30, or 31, depending on the month; the module has been programmed for 4 years (February always 28).

Day of Week: Depress the SET button a fifth time. This will display day of week. Depress the DISPLAY button and the day will advance twice each second after a 1.5 second delay. Release on the proper day.

Important: The watch will remain in any set mode with digits on until advanced to another set mode or returned to normal operation. Therefore, the setting of any function requires a complete cycle of six depressions of the SET button to return the watch to normal operation.

ALARM SETTING INSTRUCTIONS AND OPERATION:

1. Push the Alarm SET button once. On the left side a digit or digits will appear with either an "A" for A.M. or a "P" for P.M.
2. Push and hold the Alarm DISPLAY button until the desired alarm hour is reached. If you want the alarm to ring at 5:00 A.M., be sure an "A" appears. Once the correct hour is reached, release the Alarm DISPLAY button.
3. Push the Alarm SET button again and the hours previously set, and first digit of the minutes will appear.
4. Press and hold the Alarm DISPLAY button until the proper digit is reached. Note that this digit only goes to 5 and then reverts to zero since there can only be up to 50 plus minutes.
5. Push the Alarm SET button again. The second digit of the minutes will appear.
6. Press and hold the Alarm DISPLAY button until the desired second digit is reached.
7. Press the Alarm SET button once more and the screen will go blank. The alarm time is now locked in.

8. Now give the Alarm DISPLAY button two short pushes in rapid succession. The alarm time that you set will flash for 3 seconds and the screen will go black. The alarm is set to beep at the prescribed time. (*Note:* You can check to see what time the alarm is set for any time by pressing the Alarm DISPLAY button once.)
9. The alarm will sound as follows: First, a one second beep. Five seconds later, it will beep for a period of 15 seconds. A colon will appear on the screen as the alarm goes off.
10. To shut the alarm off immediately, give the Alarm DISPLAY button two short pushes.

FOR THE SLUMBER FEATURE:

If you wish to stop the alarm but have it repeat in 5 minutes, give the Alarm DISPLAY button one short push and the alarm will automatically repeat in 5 minutes. This procedure can be repeated indefinitely. If for some reason you push the Alarm DISPLAY button either accidently or to shut the alarm off, and wish to activate it again, give the Alarm DISPLAY button two short pushes and the flashing display will indicate that the alarm is set for that time. (*Note:* When the alarm is in the Slumber Setting, you cannot reset the alarm without first deactivating the alarm by pushing the Alarm DISPLAY button twice, in rapid succession.)

SIX FUNCTION: Hours, Minutes, Seconds, Month, Date, Day **LED**

TYPE: STOP WATCH

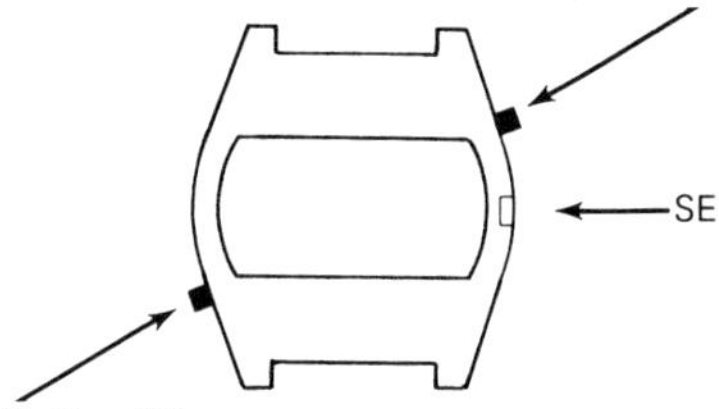

MANUFACTURERS:	RECOMMENDED BATTERY:	OTHERS:
Microsonic	UC386	______________
Windert WC-11C	UC386	______________
		______________

OPERATION:

To display hour and minutes, push DISPLAY button #1 once. Holding DISPLAY button #1 down will display seconds. To display month and date, push DISPLAY button #1 twice. Holding the button down after the second push will display the day of the week.

SETTING INSTRUCTIONS:

Month: Push the SET button once and hold it for two seconds. (*Note:* Months are shown by number.) To advance the month, depress DISPLAY button #1 and hold until the proper month is shown.

Date: To display the day of the week, depress the SET button a third time. To advance the day, depress DISPLAY button #1.

Hours: Pushing the SET button again will display the hours. To advance hours, push DISPLAY button #1.

Minutes: Depressing the SET button again displays the minutes. To advance minutes, depress DISPLAY button #1.

Seconds: For seconds, push the SET button again. Pushing the SET button once more will start the watch with seconds at zero.

STOP WATCH OPERATION:

To start, push DISPLAY button #2 once. This will show seconds, 10ths and 100ths of seconds for 1½ seconds before going blank to conserve batteries. To display elapsed time, push DISPLAY button #1 for continuous display of seconds, 10ths and 100ths. By depressing DISPLAY button #1 again, your watch will display elapsed hours and minutes. To blank out the display in order to conserve battery, push DISPLAY button #1 again.

To display correct time while Stop Watch is in use, pushing the SET button once will show you hours and minutes. Seconds will be displayed after hours and minutes, by holding the SET button. To display month and date, depress the SET button twice. By pushing the SET button twice and holding it down, you will display the day of the week.

TO STOP AND CLEAR THE STOP WATCH:

To stop the Stop Watch while it is running, push DISPLAY button #2 once. Push DISPLAY button #1 to display seconds, 10ths, and 100ths of seconds. Push DISPLAY button #1 again to display hours and minutes. By pushing DISPLAY button #1 again, display of the Stop Watch is off. Pushing DISPLAY button #2 once clears Stop Watch and gets it ready to restart.

WATCH FUNCTIONS: Hours/Minutes, Seconds, Month/Date, Day **LED**

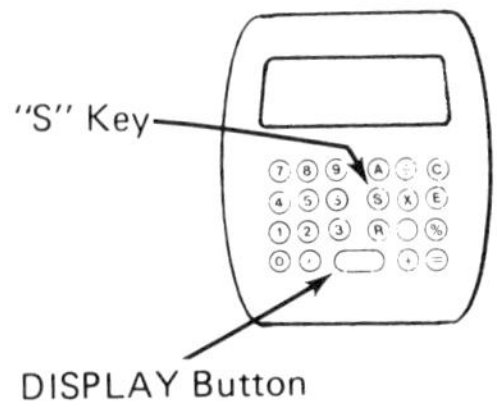

CALCULATOR FUNCTIONS: Add, Subtract, Multiply, Divide, Percent, Memory

MANUFACTURERS:	RECOMMENDED BATTERY:	OTHERS:
Pulsar	UC357	______________
		______________

WATCH OPERATION:

To display hour and minutes, press the DISPLAY button once and release. Press the DISPLAY button once and hold for at least two seconds, and seconds display follows hour/minute display, and continues for as long as you hold the button. To display month and date, press the DISPLAY button twice and release. Day of week is displayed by pressing the DISPLAY button twice and holding for at least 2 seconds.

WATCH SETTING INSTRUCTIONS:

Hours and A.M./P.M. Indicator: With the watch in normal timekeeping mode, press the SET button once for Hour and A.M./P.M. setting mode, indicated by an Hour and "A" (for A.M.) or "P" (for P.M.) on the data screen. Press and hold the DISPLAY button for at least two seconds, after which the display will advance. Release the button when hour and A.M./P.M. setting is correct. If setting hour only (ie., for time zone and daylight saving time changes), press the SET button 5 more times to return to normal timekeeping operation.

Minutes: If the watch is still in hour and A.M./P.M. setting mode, press the SET key once more for the minute setting mode. If the watch is in normal timekeeping mode, press the SET key twice for minute setting mode, indicated by a colon and two figures at right on the data screen. With watch in the minute setting mode, press and hold the

DISPLAY button for at least two seconds, after which the display will advance. Release the button when minute setting is correct. If setting minutes only, press the SET key four more times to return to normal timekeeping operation.

Seconds: To synchronize the seconds with your master time signal, first put the your watch in minute setting mode, as indicated previously. Quickly press and release the DISPLAY button once. This reactivates the seconds in coincidence with the signal. It is not necessary to advance the minutes before setting seconds. If setting seconds only, press the SET key four more times to return to normal timekeeping.

Month: If your watch is still in the minute setting mode, press the SET key once more for month setting mode. Press and hold the DISPLAY button for at least two seconds, after which the display will begin to advance. Release the button when month setting is correct.

Date: From the month setting mode, press the SET key once more for date setting mode. Press and hold the DISPLAY button for at least two seconds, after which the display will begin to advance. Release the button when Date setting is correct.

Day: Press the SET button once more for Day setting mode, indicated by two alphanumeric letters at left on the data screen. Press and hold the DISPLAY button for at least two seconds, after which the display will begin to advance. Release the button when day setting is correct. Press the SET button once more to return to normal timekeeping operation.

CALCULATOR INSTRUCTIONS:

To enter the calculate mode, press the key marked ⊕ . A zero will be displayed on the right hand digit and the calculator is now functional.

Identification of Keys:

Data Keys: The data keys include the number keys 0 through 9 and the decimal point.

Arithmetic Operation Keys:

Add	+	Multiply	×	Equals	=
Subtract	−	Divide	÷	Percent	%

Control Function Keys:

A (Add to memory) This key adds the numbers being displayed to the memory register.

S (Subtract from memory) This key subtracts the numbers being displayed from the memory register.

R (Memory Read/Clear) One push of this key will display contents of the memory register. Two successive pushes will clear the memory.

E (Clear Entry) This key clears the last number entered without affecting previous entries.

C (Clear) This key clears all previous enteries except the contents of the memory register. The memory is automatically cleared when changing from calculate mode to time mode.

Display Blanking and Power Saver: To save energy, the display will go blank after 20 seconds of keyboard inactivity. To restore display without affecting calculations, press the + or E (clear entry) key.

Calculating Mode Indicator: After 20 seconds, a calculation display will go out. To remind you that the computer is still in calculating mode, a single red dot will appear at left on the data screen.

Constant: When performing a calculation in multiply or divide mode, the multiplier or divisor can be used as a constant by successive pressings of the ×, ÷, or = keys.

Negative Number Indicator: The overflow dot to the left of the data screen also acts as a negative number indicator when the answer displayed is negative.

FUNCTIONS: Combination **LCD/LED**
Hours, Minutes, Date,
Seconds

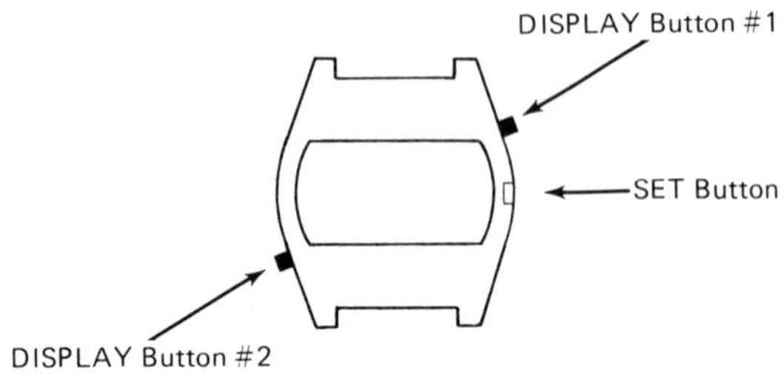

MANUFACTURERS:

Belair	Jupiter
Duratime	Longines
Elgin	Marcel
Helbros	Windert WH-4/3

OTHERS:

____
____
____
____
____

Recommended Battery: On all watches listed—UC357.

LCD OPERATION:

To display the date, simply press DISPLAY button #2. If seconds are desired, hold both DISPLAY buttons down momentarily and release. Seconds will continue to appear until either button is tapped.

LED OPERATION:

Press DISPLAY button #1 momentarily and the time in hours and minutes will appear for about 1½ seconds. To display the date, hold DISPLAY button #2 down until the LED digits light up. (For energy conservation, the LED date lags the LCD date by 1½ seconds.)

SETTING INSTRUCTIONS:

Minutes and Seconds: Depress DISPLAY button #1 and continue holding it down throughout this setting operation. Listen for the next full minute time signal. At this instant, press the SET button and hold it down until the correct minute appears. Release both buttons. Minutes and seconds are now set.

Hours: Set hours by simply pressing the SET button until the correct hour is shown and the A.M./P.M. designation is correct. The colon between hours and minutes pulsates during P.M. hours, and is continuous during A.M. hours. If the A.M./P.M. designation is not set correctly, the date will change at noon instead of midnight.

Date: While holding DISPLAY button #2 down, press the SET button until the correct date appears.

FUNCTIONS: Watch—Hours, Minutes, Seconds, Date — **LCD**

LED Stopwatch

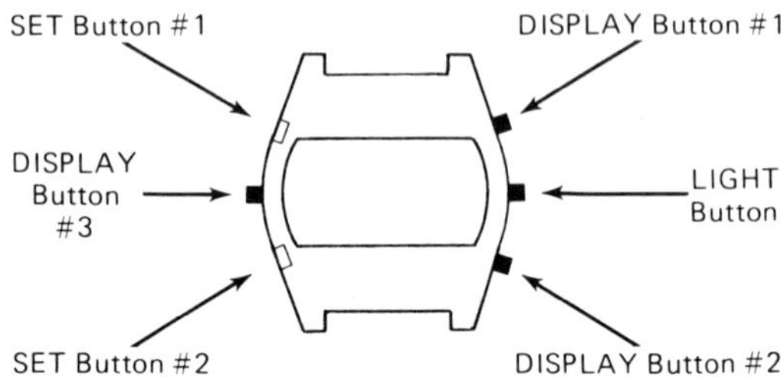

MANUFACTURERS:	RECOMMENDED BATTERY:	OTHERS:
Heuer	RW44	

LCD OPERATING INSTRUCTIONS:

The time of day display is continuously displayed on the LCD display. The control of the different LCD indications is made by DISPLAY #3 button. A depression of DISPLAY #3 button changes the display:

- from time of day to A.M./P.M. and date
- from date to seconds. After each full minute, the date changes automatically to seconds
- from seconds to time of day. After each full minute, the seconds display changes automatically back to time of day.

The flashing colon between hours and minutes (respectively on left side of seconds) indicates that the watch is running. Night time display readout is possible by means of a miniature lamp (on right side of the display) that can be switched on by depressing the LIGHT push button (on right side at 3 o'clock position).

SETTING INSTRUCTIONS:

SET button #1 advances hours if display indicates time of day or advances the A.M./P.M. indication if display indicates date. No other part of the watch is affected. The hour can be changed in this manner as you travel from one time zone to another without affecting the accuracy of the watch.

SET button #2 sets the right side of the LCD display—minutes, seconds, or date—depending on the displayed information. The watch stops when setting the minutes. After setting, display fades away (on some of the later models, only the flashing colon fades away). Restart the time of day by depressing DISPLAY #3 button, which will make LCD display reappear.

For setting seconds only, constant depressing of the DISPLAY #2 button when the seconds are displayed stops the seconds and intermittant depression accelerates counting of seconds.

STOPWATCH—FUNCTIONS IN LED:

1. Reset the stopwatch to zero through simultaneous depression of DISPLAY buttons #1 and #2. Release DISPLAY button #1 before releasing DISPLAY #2 button.
2. Verify reset by depressing simultaneously DISPLAY #3 button and DISPLAY #2 buttons. The "0" on the right side of the LED display appears as long as the buttons are held.
3. Start by depressing DISPLAY #1 button on the right side.
4. Take intermediate times once stopwatch is started by depressing DISPLAY #1 button. Each successive depression displays the corresponding intermediate time (Split-action) as long as the button is depressed plus approximately 2 seconds upon releasing it.
5. Stop by depressing DISPLAY #2 button. Total time is shown as long as the button is depressed plus approximately 2 seconds. The result is memorized, until the stopwatch is reset to zero.
6. After a stop, the display may be reset to zero as mentioned above #1, or restarted again by a new depression of DISPLAY #1 button.

Running Time: The running time can be displayed at any moment by simultaneously depressing DISPLAY #3 button and DISPLAY #1 button.

TWO FUNCTION: Hours, Minutes **LCD**

TYPE: 1

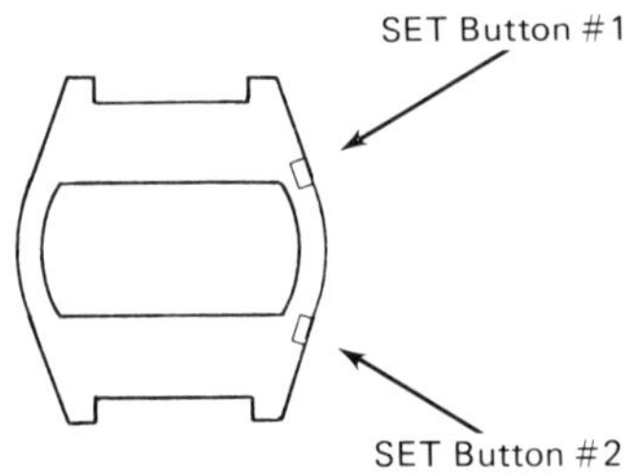

MANUFACTURERS:	RECOMMENDED BATTERY:	OTHERS:
American Microsystems Inc.	—	
Helbros Minilite	Helbros 175	
Waltham Ladies	RW37	

OPERATION:

Displays hours and minutes; colon blinks once each second.

SETTING INSTRUCTIONS:

Minutes: To advance the minutes, push both buttons simultaneously. When the correct time (in minutes) appears, release both buttons immediately.

Hours: To advance the hours, push the top button only, and hold until the correct hour is shown.

Seconds: To set to the exact second, be sure the minutes are correct. Wait five seconds and then push in and hold the bottom button only. This will advance the minute display by one minute and reset seconds to zero. Seconds will start counting when the button is released.

TWO FUNCTIONS: Hours, Minutes **LCD**

TYPE: 2

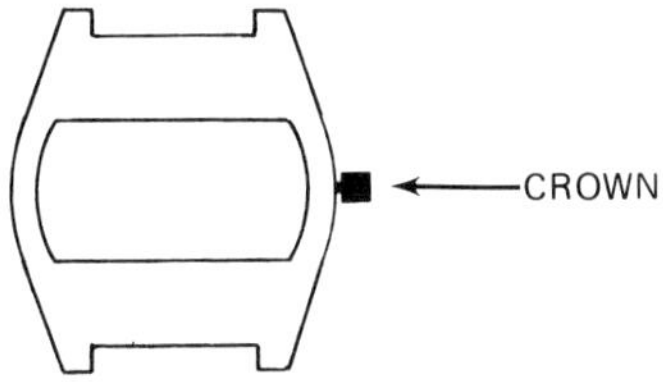

MANUFACTURERS:

American Microsystems Inc.
#HMM-717S

OTHERS:

____

____

Recommended Battery: ———

OPERATION:

Displays hours and minutes; colon blinks once each second.

SETTING INSTRUCTIONS:

Hours: Be sure the CROWN is pushed in. Turn the CROWN slightly (⅛ turn) counter clockwise. DO IT EASY! Hour digits will change at a rate of one hour per second. Turn the CROWN slightly (⅛ turn) clockwise when desired hour digits appear on display. Minutes and seconds are not affected while changing the hour digits.

Note: Do not force the CROWN past ⅛ turn.

Minutes and Seconds: Pull the CROWN out (easily done by inserting your fingernail behind the CROWN and pulling gently). In this position, the seconds are set to "0" and the watch is in a "Hold" situation. Seconds will hold at "0" until the crown is pushed in. Turn the crown slightly (¼ turn) counterclockwise. DO IT EASY! Minute digits will change at a rate of one per second. Turn the CROWN slightly (¼ turn) counterclockwise when desired minute digits appear on display. When hour and minutes digits agree with correct time, push the CROWN in.

TWO FUNCTION: Hours, Minutes **LCD**

TYPE: 3

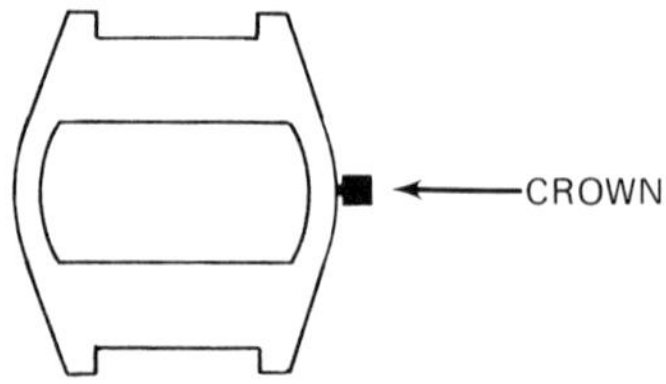

MANUFACTURERS:	**RECOMMENDED BATTERY:**	**OTHERS:**
Citizen Cal. 9030	UC384	

OPERATION:

Displays hours and minutes; colon blinks once each second.

SETTING INSTRUCTIONS:

CROWN should be in for normal operation.

Hours: Pull CROWN out to first detent for hours and minutes correction. Turn CROWN gently *clockwise* to advance hours.

Minutes: Turn CROWN gently *counterclockwise* to advance minutes.

Seconds: To synchronize seconds, pull CROWN further out to stop. Colon will stop blinking, seconds are reset to zero, and will not start until CROWN is pushed in.

TWO FUNCTION: Hours, Minutes **LCD**

TYPE: 4

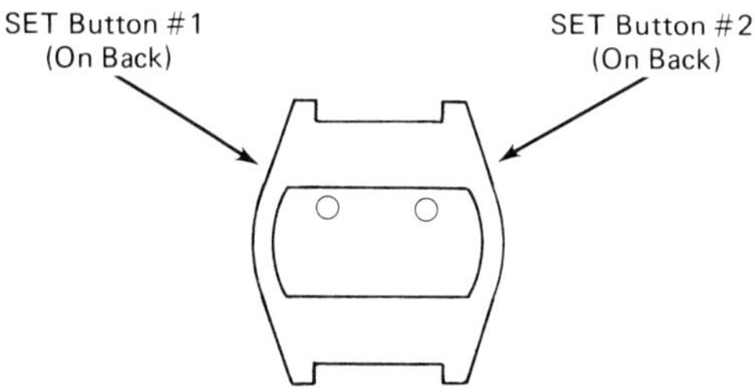

MANUFACTURERS:

Micro Display Systems "Eltek"
Sears, Roebuck Quartz Solid State

OTHERS:

____
____
____

Recommended Battery: On both models listed—UC302 or UC357

OPERATION:

The watch in its normal operation displays hours and minutes. A flashing colon (:) between the hours and minutes is used to indicate seconds.

SETTING INSTRUCTIONS:

Minutes: To set minutes, press both SET buttons on the back of the watch simultaneously. Minutes will advance at one count per second to desired minutes.

Hours: Press SET button #1 on back of watch (behind hour display) to advance hours until desired hour is reached.

Seconds: To set seconds to the exact second, set minutes one minute ahead of actual time. Depress SET button #2 (behind minutes display); at the proper minute time tone, release button and watch is set to exact second.

THREE FUNCTION: Hours, Minutes, Seconds (6 Digit) **LCD**

TYPE: 1

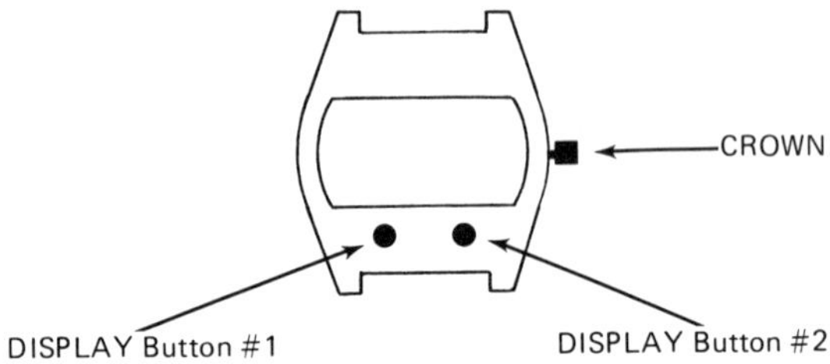

MANUFACTURERS:	RECOMMENDED BATTERY:	OTHERS:
Seiko Cal 0624	UC386	
Seiko Cal 0662	UC386	

OPERATION:

Continuous readout of hours, minutes and seconds. Push DISPLAY button #1 for display backlight.

SETTING INSTRUCTIONS:

Seconds: First, pull the CROWN all the way out to set the time. The seconds digits will immediately start flashing. When DISPLAY button #1 is then depressed in accordance with a time signal, it automatically causes the seconds digits to return to 00. (*Note:* When the seconds digits indicate 01 to 29 seconds, minutes are not changed when DISPLAY button #1 is depressed. When 30–59 seconds are displayed and DISPLAY button #1 is depressed, one minute is added.)

Minutes: While the seconds digits are flashing, depress DISPLAY button #2. The minute digits will then immediately start flashing. Each depression of DISPLAY button #1 will then advance the minute digits one minute until the desired minute display is shown.

Hours: While the minute digits are flashing, depress DISPLAY button #2 again. The hour digits will then start flashing. By depressing DISPLAY button #1, the hour digits are advanced one hour. Each depression of DISPLAY button #1 will then advance the hour digits one hour to the desired time.

Note: Be sure to push the CROWN in to the normal position after setting the time.

THREE FUNCTION: Hours, Minutes, Seconds **LCD**

TYPE: 2

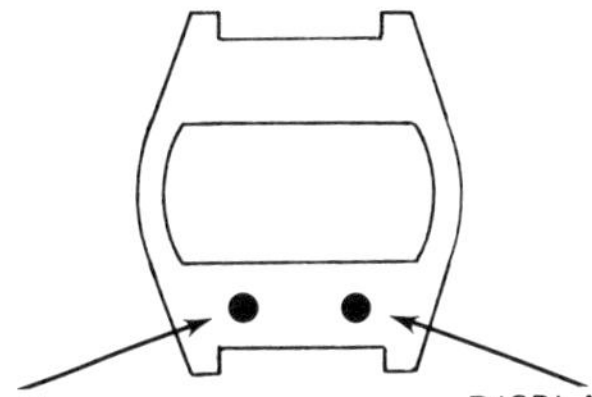

MANUFACTURERS:	**RECOMMENDED BATTERY:**	**OTHERS:**
Citizen Cal 9040	UC301	

OPERATION:

Hours and minutes are displayed continuously. The colon blinks once each second. Press the DISPLAY button to display seconds.

SETTING INSTRUCTIONS:

Seconds: Press the SET button to display seconds. Press the DISPLAY button to reset seconds to zero at time tone.

Minutes: Press the SET button a second time to display minutes. Press the DISPLAY button to advance one minute per push.

Hours: Press the SET button a third time to display hours. Press the DISPLAY button to advance one hour per push.

Press the SET button a fourth time to return watch to normal operation.

FOUR FUNCTION: Hours, Minutes, Date, Seconds **LCD**

TYPE: 1

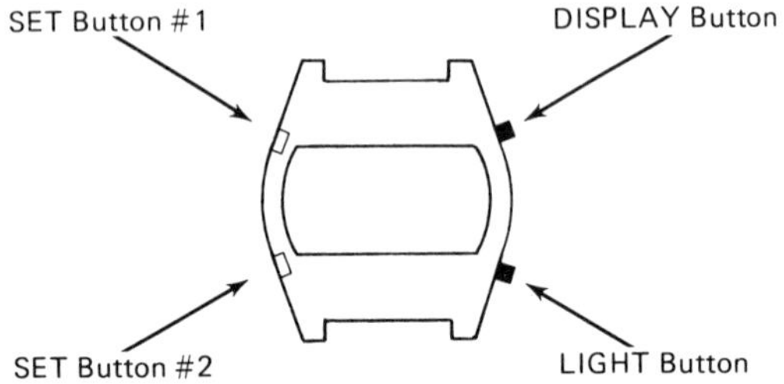

MANUFACTURERS:	**RECOMMENDED BATTERY:**	**OTHERS:**
Bulova Computron S300	RW47	
Wyler	RW47, UC392	

OPERATION:

Hours and minutes are displayed continuously. Press the DISPLAY button to display date and seconds. Press the LIGHT button to illuminate display at night.

SETTING INSTRUCTIONS:

Hours: Depress SET button #1. Release when the correct hour is shown.

Minutes/Seconds: Depress SET button #2 until the Minute shown is one minute ahead of the actual time. Release the SET button. The pulsating colon will freeze in the "on" position. The watch is now ready to be synchronized to the exact time. At the time tone, depress and release the DISPLAY button. The colon will start pulsating, and the internal counting system will begin counting the seconds.

A.M./P.M.: Depress the DISPLAY button once more and immediately release. The display will switch to the Date and the letter "A" or "P" (for A.M. and P.M.), which will appear on the left side of the display. If correction is required, depress SET button #1 until the correct letter appears. (This is necessary so that the date can be set to automatically advance at midnight and not at noon.)

Date: Depress SET button #2 until the correct date is shown.

FOUR FUNCTION: Hours, Minutes, Seconds, Date **LCD**

TYPE: 2

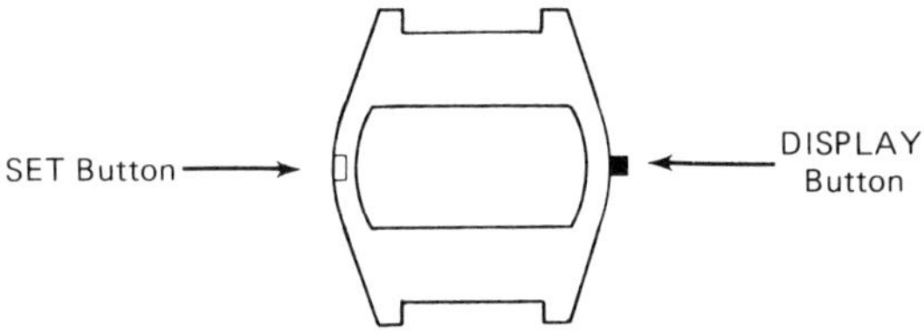

(A)

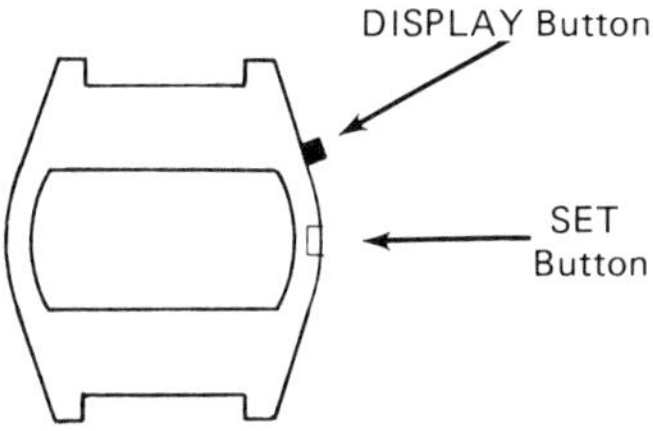

(B)

MANUFACTURERS:	RECOMMENDED BATTERY:	FIG.:	OTHERS:
Digitaire	UC309, 393 w/Light	A	________
Microma	UC393	B	________
Sensor	UC386, RW24, RW44	B	________
Sears, Roebuck	UC386	B	________

OPERATION:

When the DISPLAY button is depressed, display will show seconds. Next push will cause date to be displayed, and a third push will return the display to hours and minutes. The colon flashes at a 1 flash per second rate except in the setting mode.

SETTING INSTRUCTIONS:

Minutes: Push and release the SET button to display minutes only. In this mode, the minutes can be advanced by depressing the DISPLAY button. Set minutes one minute *behind* intended set time.

Seconds: The next push and release of the SET button will cause seconds to be displayed. Depressing the DISPLAY button will cause seconds to be set to zero and held until the button is released. Minutes will also be advanced by one minute.

Hours: The third push and release of the SET button will cause hours to be displayed. Colon is present to indicate P.M. Hours are advanced when the DISPLAY button is advanced.

Date: The fourth push and release of the SET button will display date and setting is accomplished the same as were hours and minutes.

The fifth push and release of the SET button will return the watch to normal operation.

FOUR FUNCTION: Hours, Minutes, Date, Seconds **LCD**

TYPE: 3

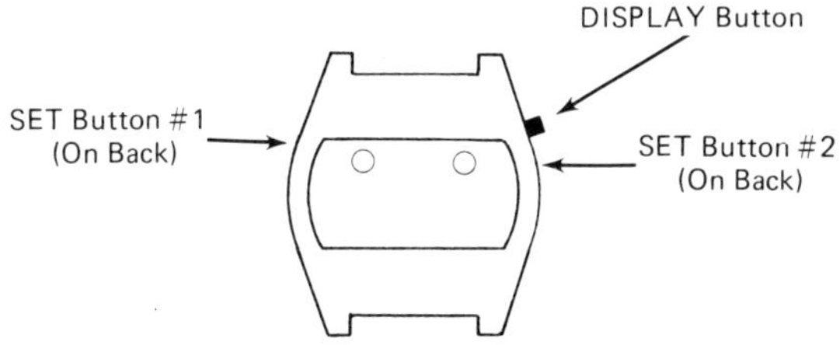

MANUFACTURERS:	**RECOMMENDED BATTERY:**	**OTHERS:**
Sensor	UC303 or 357	____________
Sears, Roebuck and Co.	UC303 or 357	____________
Baylor	UC303 or 357	____________

Note: Button to illuminate display located at 4 o'clock on some models.

OPERATION:

Constant display of hours and minutes. Colon remains constant in A.M., flashes each second in P.M. Push the DISPLAY button for date (at place of hours) and seconds (at place of minutes), no colon. Some models built to give 24 hour display instead of 12 hours. Some models have backlight turned on by depressing LIGHT button.

SETTING INSTRUCTIONS:

Hours/minutes: Hours and minutes are advanced at 1 Hz by depressing SET button #1 or #2 behind appropriate display.

Date/seconds: Date advanced at 1 Hz and seconds set to zero by depressing the DISPLAY button to call these displays, then depressing appropriate SET button on back of watch.

Note: All settings are independent; that is, advancing minutes, for example, to 59 with the SET button does not advance hours, etc.

FOUR FUNCTION: Hours, Minutes, Seconds OR Hours, Minutes, Date **LCD**

TYPE: 4

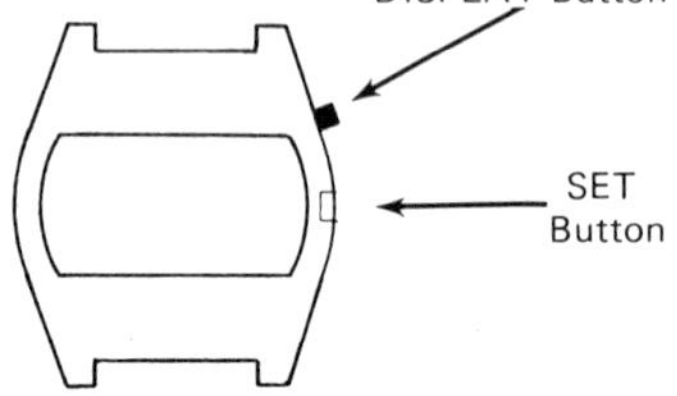

MANUFACTURERS:	RECOMMENDED BATTERY	OTHERS:
Savant	For all models:	__________
Sears, Roebuck and Co.	UC393 (w/backlight)	__________
Sensor	UC392 (w/"TriLite"®)	__________

Note: Button to illuminate display located at 4 o'clock on some models.

OPERATION:

Hours and minutes are constantly displayed, and the display may be changed from seconds to date or from date to seconds by simply depressing the DISPLAY button. For night viewing, press and hold the LIGHT button. The recessed button between the DISPLAY button and the LIGHT button is for setting the various functions.

SETTING INSTRUCTIONS:

Hours: Press the SET button once to show only hour as well as an "A" or a "P" to indicate A.M./P.M. Press the DISPLAY button and hold until correct hour is indicated.

Minutes: Press the SET button once again to show minutes only. Set by holding down the DISPLAY button. Set minutes one minute ahead of the actual time. Setting minutes freezes the seconds at zero.

Month: Press the SET button a third time to show month. Although the month does not appear on the normal display, it must be set so the watch can automatically change the days each month. Holding down the DISPLAY button will change months.

Date: Press the SET button a fourth time to show the date, which can be set by depressing the DISPLAY button.

To start watch functioning, press the SET button a fifth time to return display to hours, minutes and seconds. If the minutes had been set, the seconds will be holding at 00. At the time signal indicating the minute to which watch was set as above, depress the DISPLAY button to start watch counting from 00 seconds. If minutes had not been changed, watch would be in normal operating mode after fifth pressing of the SET button.

FIVE FUNCTION: Hours, Minutes, Day, Seconds, Or Date **LCD**

TYPE: 1

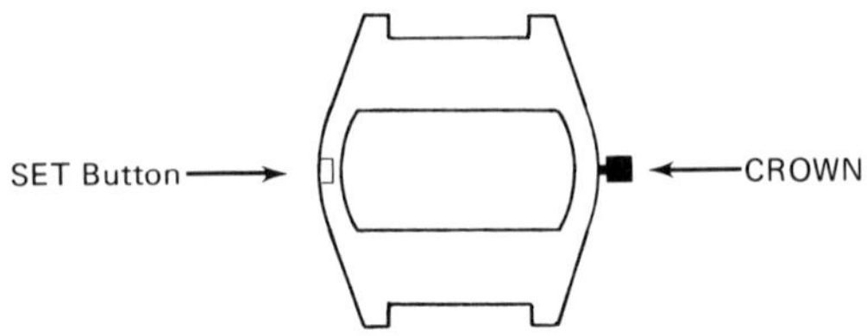

MANUFACTURERS:	**RECOMMENDED BATTERY:**	**OTHERS:**
Citizens Cal 9010	UC301	
Citizens Cal 9011	UC301	
Citizens Cal 9012	UC301	

OPERATION:

Displays hour, minutes, A.M. or P.M., day and date. Press the CROWN in to display seconds in place of date.

SETTING INSTRUCTIONS:

Minutes: Pull the CROWN past first detent to the "Out" position. Gently turn counterclockwise to advance minutes.

Hours: Same procedure as for minutes, except the CROWN should be gently turned clockwise to advance hours.

Date: Pull the CROWN to first detent. Gently turn counterclockwise to advance date.

Day: Same procedure as for date, except the CROWN should be gently turned clockwise to advance day.

Seconds: Press the SET button to zero seconds.

Note: Power to watch can be turned "Off" by rotating the CROWN clockwise while in the normal "In" position.

FIVE FUNCTION: Hours, Minutees, Month, Date, Seconds **LCD**

TYPE: 2

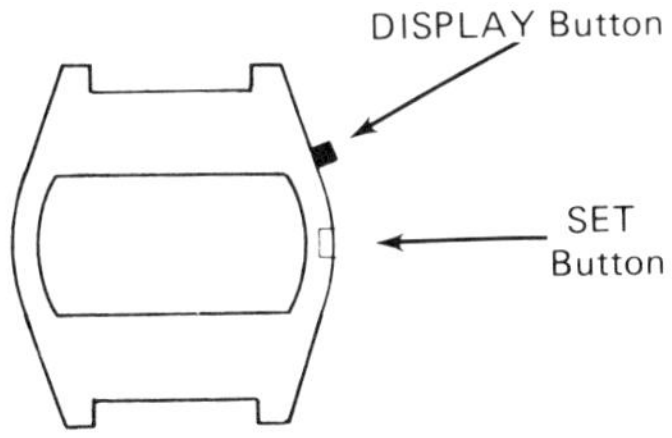

MANUFACTURERS:	RECOMMENDED BATTERY:	OTHERS:
Fairchild	RW48, UC393, M10L123	
Novus	RW48, UC393, M10L123	
Sears Roebuck "Phasar"	Various battery thickness	
Sensor/Savant	Various battery thickness	
Speidel	Mens—UC393, RW48; Ladies—UC392, RW37	
Timeband	RW48, UC393, M10L123	
Waltham	Mens—UC393, RW48; Ladies—RW47	

Note: Button to illuminate display located at 4 o'clock on some models.

OPERATION:

When the DISPLAY button is depressed, month/date will be displayed for 2–3 seconds and then automatically will return to hours/minutes. Seconds can be displayed by depressing the DISPLAY button twice in rapid succession and will remain on until DISPLAY button is depressed again returning display to hours/minutes. On some models, depressing the SET button with ballpoint pen while display is on hours - minutes will cause watch to alternately show hours/minutes, month/date.

SETTING INSTRUCTIONS:

Month: Press the SET button once to set month. Depressing the DISPLAY button will advance month at one month per second.

Date: Press the SET button again to set date. Depressing the DISPLAY button will advance date at one day per second.

Hours: Again depress the SET button. Depressing the DISPLAY button will advance hours at one hour per second. A.M. or P.M. is indicated by an "A" or "P."

Minutes: Again press the SET button. Depressing the DISPLAY button will advance minutes at one minute per second. Set minutes one minute ahead and then press the SET button again to return display to hours/minutes. Seconds counter is holding at zero.

At time signal indicating minute to which watch is being set, press the DISPLAY button, which starts seconds counter. The watch is now in normal operation and set to exact seconds.

FIVE FUNCTION: Hours, Minutes, Day, Date with Bar showing Seconds **LCD**

TYPE: 3

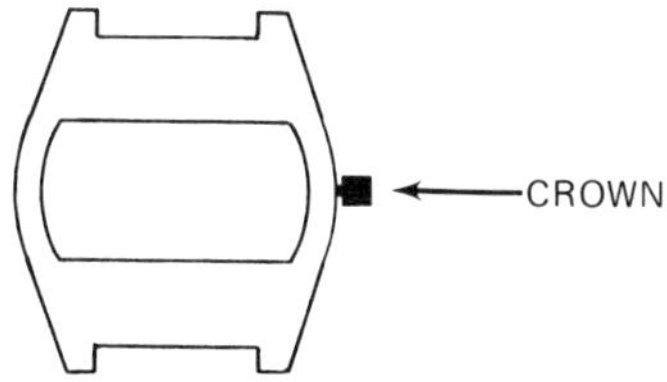

MANUFACTURERS:	RECOMMENDED BATTERY:	OTHERS:
Seiko Cal. 0124	UC386	

OPERATION:

Continuously shows hours, minutes, day and date. A series of six lines between the time and day display flash to show seconds. Each bar turned on indicating 10 seconds.

SETTING INSTRUCTIONS:

Date: First, pull the CROWN all the way, and then turn the CROWN clockwise to indicate only the date and seconds. The date is changed when the CROWN is turned counterclockwise. When the desired date appears, return the CROWN to the center position. Leave the CROWN in the pulled out position.

Day of Week: By next turning the CROWN clockwise, the day of the week and the seconds are indicated and the other indications are extinguished. The day of the week is indicated in the square frame. The day of the week is changed when the CROWN is turned counterclockwise. When the desired day appears, return the CROWN to the center position. Leave the CROWN in the pulled out position.

Hours: By next turning the CROWN clockwise, only the hour and seconds are indicated. The hour is changed when the CROWN is turned counterclockwise. When the desired time appears, return the CROWN to the center position. (*Note:* When setting the hours, A.M. and P.M. indications are also changed automatically. Be sure that the time setting is made with A.M. or P.M. appropriately displayed.)

Minutes/Seconds: By next turning the CROWN clockwise, the minutes and seconds are indicated. The minutes are changed successively by turning the CROWN counterclockwise. When this is done, seconds readout is returned to zero and stops counting. When the desired minute appears, return the CROWN to the center position. The second readout is started exactly from zero when the CROWN is pushed back to the normal position in accordance with a time signal. The date, day, hour, minute, and seconds are also displayed at the same time.

SIX FUNCTION: Hours, Minutes, Day, Date, Month, Seconds **LCD**

TYPE: 1

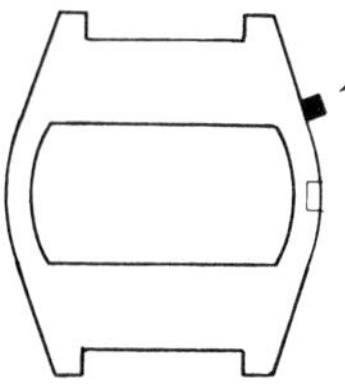

MANUFACTURERS:	**RECOMMENDED BATTERY:**	**OTHERS:**
Sensor	RW49, M10L122	

Note: Button to illuminate display located at 4 o'clock on some models.

OPERATION:

Displays hours and minutes with blinking colon to indicate watch is operating. One push of the DISPLAY button will change the display from hours and minutes to the day of the week and date, and then will change again to show the month, such as 11 for November. The display automatically returns to hours and minutes. Pushing the DISPLAY button twice rapidly will change the display to read seconds and will continue to count until the DISPLAY button is pushed again. For night viewing, press and hold the LIGHT button. This watch may be set to automatically change the display from hours and minutes to day and date by pressing the SET button once. To return display to hours and minutes requires running the watch through the setting steps listed below.

SETTING INSTRUCTIONS:

Depressing the SET button while display is on hours/minutes will alternately show hours/minutes, day/date.

Month: Depress the SET button second time to set month. Depressing the DISPLAY button will advance month at one month per second.

Date: Press the SET button a third time to set date. Depressing the DISPLAY button will advance date at one day per second.

Day: Press the SET button a fourth time to set day. Depressing the DISPLAY button will advance day at one day per second.

Hours: Press the SET button a fifth time to set hours. Depressing the DISPLAY button will advance hours at one hour per second. A.M. or P.M. is indicated by an "A" or "P."

Minutes: Press the SET button a sixth time to set minutes. Depressing the DISPLAY button will advance minutes at one minute per second. Set minutes one minute ahead and press SET button to return display to hours/minutes. Seconds counter will hold at zero.

At time signal indicating minute to which watch is set, press the DISPLAY button, which starts seconds counter, and watch is now in normal operation.

CHRONOGRAPH: WATCH FUNCTIONS— Hours, Minutes, Date **LCD**

CHRONOGRAPH FUNCTIONS— Minutes, Seconds, 1/10 Seconds, Elapsed Time and Lap Time

TYPE: 1

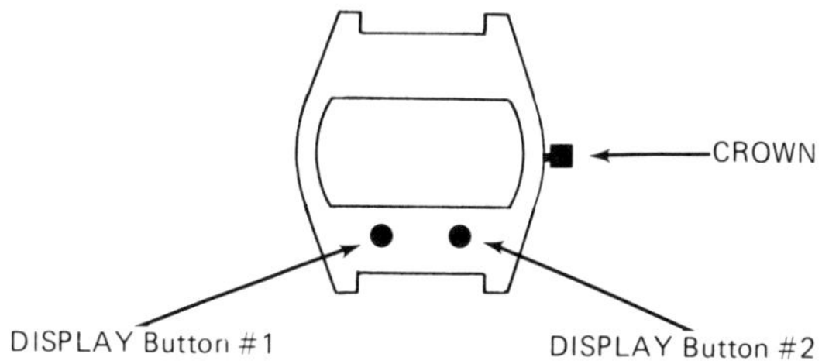

MANUFACTURERS:	RECOMMENDED BATTERY:	OTHERS:
Seiko Cal. 0634	UC386	______
		______
		______

OPERATION:

Displays hours, minutes and date in watch operation. DISPLAY button #1 can also be pressed to illuminate display.

SETTING INSTRUCTIONS:

Seconds: First, pull the CROWN, all the way out to set the time. When DISPLAY button #1 is pushed, in accordance with a time signal, it automatically causes the seconds to return to 00.

Date: With the CROWN pulled all the way out, push DISPLAY button #2 and the date digits will immediately start flashing. By pushing DISPLAY button #1, the date digits are advanced one digit. Continue to push until the desired date is reached.

Hours: While the date digits are flashing, push DISPLAY button #2 again, and the hour digits will start flashing. By depressing DISPLAY button #1, the hour digits are advanced one hour. Each depression of DISPLAY button #1 will then advance the hour digits one hour to the desired time. Be sure that the time setting is made with A.M. or P.M. appropriately displayed.

Minutes: While the hour digits are flashing, push DISPLAY button #2. The minutes digits will then immediately start flashing. Each depression of DISPLAY button #1 will then advance the minute digits one minute until the desired minute display is shown.

Be sure to push the CROWN in to the normal position each time after you have finished setting the time. When the CROWN is at the normal position, the time display is not changed, even if DISPLAY buttons #1 or #2 are pushed when the watch is worn on a wrist.

CHRONOGRAPH INSTRUCTIONS:

When the watch is indicating the time, the time display is instantly changed to the chronograph display by pushing the CROWN in. DISPLAY buttons #1 and #2 then function as the chronograph operation buttons.

1. To Start and Stop Timing—After pushing in the CROWN, push DISPLAY button #2 to start chronograph operation; push it again to stop and push DISPLAY button #1 to reset the chronograph display back to zero. So long as DISPLAY button #2 is pushed repeatedly, it is possible to accumulate the time being measured.

2. To Measure Elapsed Time—Push DISPLAY button #2 to start chronograph operation; and push DISPLAY button #1 when elapsed time is to be measured. The time display stops and LAP appears in the upper left corner of the display after DISPLAY button #1 is depressed. The time display then indicates the elapsed time. However, the second flashing mark continues to blink, indicating that the chronograph mechanism is still in operation. If DISPLAY button #1 is pushed, while the static elapsed time is displayed, the display automatically returns to show the time being measured from when the chronograph function was first started.

Pushing DISPLAY button #2 stops the chronograph operation, and pushing DISPLAY button #1 causes the reset to return the indication to zero. If DISPLAY button #2 is pushed during the measurement of an elapsed time, the chronograph operation stops, showing the static elapsed time. If DISPLAY button #1 is pushed in this condition, the elapsed time display is extinguished and the watch displays the measured time from the start of the chronograph function up to the moment that DISPLAY button #2 was pushed. Pushing DISPLAY button #1 again resets the watch and indicates zero.

Note: Pushing the CROWN during the chronograph operation will change the display to show the date and time, and pushing the CROWN again will change the display to again indicate the time being measured by the chronograph function. There is no confusion between timekeeping and the chronograph device as each functions independently. When the chronograph device is to be used for a long time, it is recommended that the display be changed into the timekeeping mode (by pushing the CROWN in) to prevent pushing DISPLAY buttons #1 and #2 by mistake during the time measurement. The CROWN can then be pushed in again when the time measurement is required.

CHRONOGRAPH: WATCH FUNCTIONS— **LCD**
Hours, Minutes, Date,
or Hours, Minutes, Seconds

CHRONOGRAPH FUNCTIONS—
Minutes, Seconds, 1/10
Seconds, Elapsed Time,
Lap Time

TYPE: 2

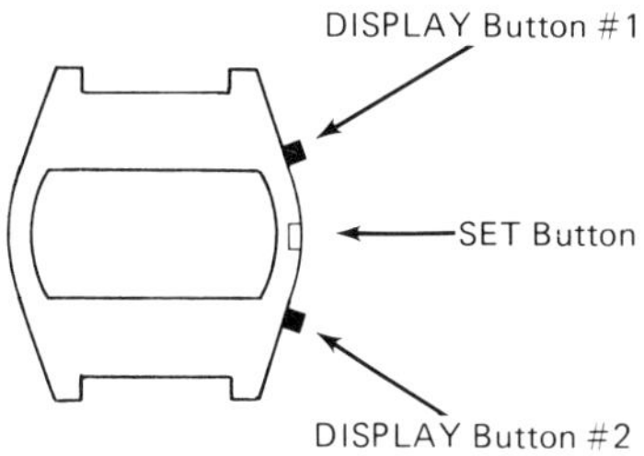

MANUFACTURERS:	RECOMMENDED BATTERY:	OTHERS:
Nepro	UC392	
Savant	UC392, M10L125	
Sensor	UC392, M10L125	

OPERATION:

Hours and minutes are on constant display and you may change from seconds to date or from date to seconds by pressing DISPLAY button #1. To switch from time mode to chronograph, press DISPLAY button #2 twice to show all zeros. Watch will continue to record time while in chronograph mode.

SETTING INSTRUCTIONS:

Hours: Press the SET button once to show only hours, as well as an "A" or "P" to indicate A.M./P.M. Press DISPLAY button #1 and hold until correct hour is reached.

Minutes: Press the SET button again to show minutes and seconds. At time signal, press DISPLAY button #1 to zero seconds. If the seconds count was between 0 and 29, the minutes will stay the same. If seconds count was between 30 and 59, the minutes will advance one minute.

Month: Press the SET button a third time to show month. Although the month does not appear on the normal display, it must be set to automatically change at the proper date each month. Holding down DISPLAY button #1 will change months.

Date: Press the SET button a fourth time to show the date, which can be set by holding down DISPLAY button #1.

To return to timekeeping mode, press the SET button a fifth time to return display to hours and minutes and seconds.

CHRONOGRAPH INSTRUCTIONS:

To switch to chronograph mode, depress the DISPLAY button two times, at which time the display will show all zeros.

Chronograph with 1/10th second timing and time-freeze feature:

1. To start timing, push DISPLAY button #1 once.

2. To stop timing, push DISPLAY button #1 again. Display will show elapsed time to 1/10th second.

3. To return to standard timekeeping mode, push DISPLAY button #2 twice within one second.

Lap Timing Mode:

1. While in chronograph mode, push DISPLAY button #1 once to start timing.

2. To freeze lap time, push DISPLAY button #2 once. Display shows lap time. Push DISPLAY button #2 again and chronograph returns to running total. Procedure may be repeated as often as desired.

3. To stop timing, push DISPLAY button #1 once. Display shows total elapsed time.

4. To return to standard timekeeping mode, push DISPLAY button #2 twice within one second.

CHRONOGRAPH: WATCH FUNCTIONS— **LCD**
Hours, Minutes,
Seconds, Month,
Date, Day

CHRONOGRAPH FUNCTIONS—
Minutes, Seconds, 1/10th
and 1/100ths Seconds,
Elapsed Time, Lap Time

TYPE: 3

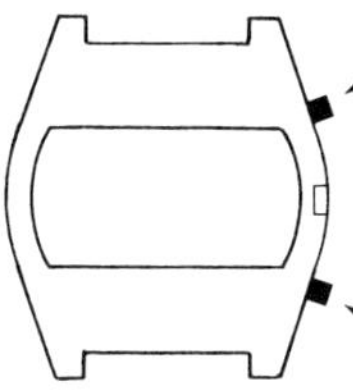

MANUFACTURERS:	**RECOMMENDED BATTERY:**	**OTHERS:**
Savant	UC392, M10L125	______
Sensor	UC392, M10L125	______
		______

OPERATION:

Hours and minutes and seconds are on constant display. Push and hold DISPLAY button #1 to show month, date and day of week. Release button and display returns to hours, minutes and seconds.

To switch from time mode to chronograph mode, push and hold DISPLAY button #2 until all zeros show. To return to timekeeping mode, push and hold DISPLAY button #2 until time shows. Star on display during timekeeping indicates P.M. The same star during chronograph mode indicates lap time.

SETTING INSTRUCTIONS:

During the setting procedure, the blinking digits indicate what function is being set. During setting of digits: (1) Pushing the DISPLAY button #1 will stop digits from advancing and will return display to timekeeping mode. (2) Pushing the SET button will stop digits from advancing but will cause next function to be set to blink and advance. Setting sequences is seconds, hours, first digit of minutes, second digit of minutes, and time freeze.

Seconds: Press the SET button to show hours and minutes and blinking seconds. To zero seconds at the time tone, depress DISPLAY button #1. If the seconds count was between 0 and 29, the minutes will stay the same; if seconds count was between 30 and 59, the minute will advance one minute.

Hour/Minutes: To set the hour or either digit of the minutes, press the SET button until desired digit is blinking and when correct number is showing, depress DISPLAY button #1 or SET button to freeze correct digit.

Pushing the SET button after setting minutes will freeze display with seconds holding at zero. Push either DISPLAY button #1 or the SET button to start time counting.

Month/Date/Day: Push both DISPLAY button #1 and the SET button at the same time until blinking occurs. Setting of these functions is accomplished by following the same procedure outlined above.

CHRONOGRAPH INSTRUCTIONS:

To switch to chronograph mode, push DISPLAY button #2 until display shows all zeros.

Chronograph with 1/100th Second timing and time-freeze feature:

1. To start timing, push DISPLAY button #1 once.
2. To stop timing, push DISPLAY button #1 again. Display will show to 1/100th second.
3. To return to standard timekeeping mode, push and hold DISPLAY button #2.

Lap Timing Mode:

1. While in the chronograph mode, push DISPLAY button #1 once.
2. To freeze lap time, push DISPLAY button #2 once. Display shows lap time. Push DISPLAY button #2 again and chronograph returns to running total. Procedure may be repeated as often as desired.
3. To stop timing, push DISPLAY button #1 once. Display shows total elapsed time.
4. To return to standard timekeeping mode, push and hold DISPLAY button #1.

ALARM: WATCH FUNCTIONS— Hours, Minutes, Seconds **LCD**

ALARM SETTING—
Alarm set to 1 minute up to 12 hours.

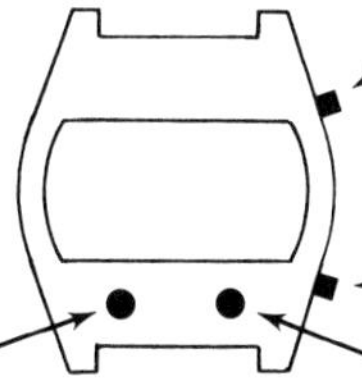

MANUFACTURERS:	**RECOMMENDED BATTERY:**	**OTHERS:**
Citizen #9021	UC386	

OPERATION:

Hours, minutes and seconds are displayed continuously. Press the LIGHT button to illuminate display for night viewing.

WATCH SETTING INSTRUCTIONS:

Seconds: Press the SET button to cause seconds to blink. Push DISPLAY button #1 at time tone to zero seconds.

Minutes: Press the SET button a second time to cause minutes to blink. Minutes will advance one minute per push of DISPLAY button #1.

Hours: Press the SET button a third time to cause hours to blink. Hours will advance one hour per push of DISPLAY button #1.

Press the SET button a fourth time to return to normal operation.

ALARM SETTING INSTRUCTIONS:

Press DISPLAY button #1 to go to alarm set mode. Seconds will disappear and a small square will appear to indicate alarm mode. Time of alarm is indicated in this mode. To change time of alarm, depress DISPLAY button #1 a second time and hours will start to blink. Hours will be advanced one hour per push of DISPLAY button #2. To change minutes setting (alarm time), press DISPLAY button #1 a third time. Minutes will be advanced one minute per push of DISPLAY button #2. When correct alarm time is set, press DISPLAY button #1 a fourth time to return to normal operating mode.

ALARM OPERATION:

To turn alarm set on and off, first depress DISPLAY button #1. Alarm time will be displayed. No display in the seconds position indicates alarm is off—it will not buzz at the indicated time. A square in the second display position indicates the alarm is on and will sound the buzzer at the set time.

In this mode, alarm can be changed from on to off, off to on, etc., by depressing the SET button.

Depress DISPLAY button #1 to go back to normal timekeeping. When alarm sounds, it can be stopped by depressing DISPLAY button #2.

INDEX

APPENDIX B
INDEX

LED WATCHES

LED WATCHES

LED WATCHES

LED WATCHES

COMBINATION LED AND LCD

LCD WATCHES